Eco-Novel Food and Feed

Eco-Novel Food and Feed

Editors

Isabel Sousa
Anabela Raymundo
María Dolores Torres

MDPI • Basel • Beijing • Wuhan • Barcelona • Belgrade • Manchester • Tokyo • Cluj • Tianjin

Editors
Isabel Sousa
Universidade de Lisboa
Portugal

Anabela Raymundo
Universidade de Lisboa
Portugal

María Dolores Torres
Universidade de Vigo
Spain

Editorial Office
MDPI
St. Alban-Anlage 66
4052 Basel, Switzerland

This is a reprint of articles from the Special Issue published online in the open access journal *Applied Sciences* (ISSN 2076-3417) (available at: https://www.mdpi.com/journal/applsci/special_issues/Eco-Novel_Food_Feed).

For citation purposes, cite each article independently as indicated on the article page online and as indicated below:

LastName, A.A.; LastName, B.B.; LastName, C.C. Article Title. *Journal Name* **Year**, *Article Number*, Page Range.

ISBN 978-3-03943-108-3 (Hbk)
ISBN 978-3-03943-109-0 (PDF)

Cover image courtesy of Isabel Sousa.

Contents

About the Editors

Isabel Sousa is Associate Professor with Habilitation at the University of Lisbon, graduated in Food Engineering and Ph.D. in Food Science by the University of Nottingham, UK. Head of LEAF research centre. Pioneered the studies on Food Texture and Rheology in Portugal and set up the Food Rheology and Cereal Technology Labs to support research in these areas. Focused in sustainability and efficiency, is involved in projects with the Industry, mainly in boosting Innovation through product development e.g. using functional ingredients, upcycling by-products and underexplored food sources, into vegetable staple foods, with strong impact on consumer's wellbeing. International Expert evaluator (since 2005) for: i) European Commission, namely for the Research Executive Agency (REA) of EC; ii) the Eurostar program of the Eureka Secretariat; and iii) the Innovation Fund Denmark (IFD).

Anabela Raymundo is Chemical engineer, MSc in Food Science and Technology and a Ph.D. in Food Engineering. Assistant Professor with Habilitation, integrating the LEAF (Linking Landscape, Environment, Agriculture and Food) research centre. Responsible for the areas of Rheology and Food Texture and Quality Control in Food Engineering Graduation and Master and in the Master of Gastronomic Sciences. Main work focused on the use of poorly exploited food sources (e.g., microalgae biomass and food industry by-products) for the development of high added value products.

María Dolores Torres was awarded her Ph.D. in Chemical and Environmental Engineering at the University of Santiago de Compostela. She has a research position at the University of Vigo (Spain), where is working on the integral valorisation of by-products from the agri-food industry and the development of innovative functional gelled foodstuff, mainly directed towards target population groups with special nutritional requirements. Currently working on the integral valorisation of brown and red algae using environmental friendly technologies to obtain functional hydrogels and other components of industrial value.

Preface to "Eco-Novel Food and Feed"

To keep up with the dynamics of the food business, the food industry has for a permanent need to develop new food products, adjusting to consumer demands and, in the near future, to the scarcity of food resources and sustainability boundaries. With the expectation of a demographic burst from the current 7.8 billion up to 9 billion in less than three decades, the production of food—especially proteins—must be increased by about 70% to meet the population nutrition needs. The concepts of sustainable food production, food products as health and wellness promoters, the use of alternative ingredients such as new protein sources, and the use of by-products in designing food or feed formulations according to bioeconomic principles are current topics that act as driving forces for innovation. With the increasing awareness of our endangered planet, the finitude of resources, and climate change, sustainability is coming to the forefront of human conciousness and activities. Sustainability in the production of food ingredients and the economic viability of their production, as well as their subsequent transformation into well-accepted commercial food products, are essential for the progress of the food industry. These have a strong impact on the economy and wellbeing. The use of food industry by-products as a source of food ingredients (e.g., proteins, structuring biopolymers, fibers) along with underexplored sources of food (e.g., macro or microalgae, psyllium, insects) are some of the challenges in creating novel food or feed products for large or niche markets, such as vegan foods, gluten-free foods, salt- or sugar-free foods, etc. Finally, the consumer attitude towards new food products is a relevant issue for the success of the novelties, and should be assessed for close-to-market novel products. Innovation, eco-friendliness, and economics are the words for success of Eco-Novel Foods and Feed.

Isabel Sousa , Anabela Raymundo, María Dolores Torres
Editors

Editorial

Special Issue: Eco-Novel Food and Feed

Isabel Sousa [1,*], Anabela Raymundo [1,*] and María Dolores Torres [2,*]

[1] LEAF—Linking Landscape, Environment, Agriculture and Food, Instituto Superior de Agronomia, Universidade de Lisboa, Tapada da Ajuda, 1349-017 Lisbon, Portugal
[2] Department of Chemical Engineering, Science Faculty, Universidade de Vigo (Campus Ourense), As Lagoas, 32004 Ourense, Spain
* Correspondence: isabelsousa@isa.ulisboa.pt (I.S.); anabraymundo@isa.ulisboa.pt (A.R.); matorres@uvigo.es (M.D.T.)

Received: 18 June 2020; Accepted: 25 June 2020; Published: 5 July 2020

Abstract: Tendencies in food and feed industries deal with a permanent need to develop innovative products, tailored to consumer demands and, in the near future, to scarcity of food resources. Sustainable food production and food products as health and wellness promoters, and the use of alternative ingredients or by-products in designed thought food or feed formulations following circular economy principles, are hot topics that act as driving forces for innovation. This special issue offers a comprehensive forum for exchanging novel research ideas or empirical practices covering discussions from healthy foodstuffs enriched with functional ingredients, with special emphasis on those targets for populations with specific requirements, to consumer attitudes towards new ingredients and end products.

Keywords: consumer acceptance and attitude; food for groups with special requirements; health and wellness promoters; innovation; product development; sustainability and economic viability

---

1. Introduction

Nowadays, the development of healthy foods enhanced with functional ingredients rose considerably as the relation between diet and health became a priority for consumers. Sustainability in the production of food or feed ingredients and economic viability of their production and subsequent transformation into commercially well-accepted final products are dramatically relevant for the progress of the industry with an essential role on the economy. The methodologies for food or feed product development are currently based on the chemical, nutritional and mechanical features, accompanied with a sensorial analysis of the final product. Moreover, the structure of foodstuffs is demonstrated to be decisive for food appeal and strongly impacts consumers' acceptance. In the creative process, the food biopolymers are the major players for the creation of relevant food structures such as foams, emulsions or gels. The development of products tailored to the needs of specific target groups like people with food intolerances, babies, elderly people, athletes or even astronauts, who often require appropriate nutritional solutions, by using biopolymers, as well as the use of food industry by-products as source of these structuring macromolecules, along with the structural implications of adding novel ingredients, are some of the challenges in creating novel food or feed products. Lastly, consumer attitude towards new products is a critically relevant concern for the accomplishment of the novelties and should be considered for close to market novel products.

2. Contributions

The papers included in this special issue cover discussions on: Food neophobia or distrust of novelties?; Exploring consumers' attitudes toward GMOs, insects and cultured meat; Increased grain amino acid content in rice with earthworm castings; Antihypertensive peptide activity in

Dutch-type cheese models prepared with different additional strains of *Lactobacillus* genus bacteria; Increased anti-inflammatory effects on LPS-induced microglia cells by *Spirulina maxima* extract from ultrasonic process; A novel way for whey: Cheese whey fermentation produces an effective and environmentally-safe alternative to chlorine; Wheat bread with dairy products technology, nutritional and sensory properties; or *Psyllium* and *Laminaria* partnership—an overview of possible food gel applications. The editors acknowledge all contributions, and we are delighted to introduce a collection of seven selected high-quality research papers in this Special Issue.

Facio and Fovino [1] presented a comprehensive review on the challenges the food industry continually faces to find new ideas to satisfy the increasingly specific consumer demand, since innovative food products do not always become part of consumption habits or create a real market. They stated that one of the major sources of resistance to novelty lies in the attitude of the consumer, who in many cases may be suspicious or hostile as a result of specific ideologies, overly attached to tradition or affected by neophobia, which was discussed throughout the paper. Moreover, they reviewed the recent literature on Europeans' attitude toward novel foods and innovation, including genetically modified organisms, cultivated meat and insects as food, which revealed a number of paradoxes in consumers' behavior, and in the many complex conditions underpinning the success of innovation in food production. Consumer involvement in the early stages of the development process is one of the strategies aimed at minimizing the failure of new products when they reach the market.

Huang and coworkers [2] contributed with a brief report on the nutritional value of rice to enhance the health of rice consumers, indicating that grain amino acid content is an important nutritional component. This study was conducted to test the hypothesis that the application of earthworm castings could increase the grain amino acid content in rice. Their outcomes showed that total amino acid content in the grain was significantly elevated by applying earthworm castings, with an average increase of 8% across four tested rice cultivars. These authors suggested that this behavior can be related to improving the efficiency of the nitrogen to amino acid conversion, and highlighted that further studies are required to assess the effects of earthworm castings on the amino acid metabolism in rice grains.

Garbowska and coworkers [3] focused on the proteolytic activity of bacterial strains from the genus *Lactobacillus* and their capability in producing peptide inhibitors of angiotensin-converting enzyme in cheese models prepared with their addition. These authors indicated that all tested cheese models exhibited a high ability of angiotensin convertase inhibition (>80%, after five weeks of ripening). They also found that use of the adjunct bacterial cultures from the genus *Lactobacillus* contributed to lower IC_{50}. In addition, they pointed out that the proteolytic activity of model cheeses varied in their increase through the period of ripening, with changes in values dependent on the adjunct lactic acid bacteria strain used for cheese making.

Choi and coworkers [4] assessed the anti-inflammatory impact of *Spirulina* extract from a non-thermal ultrasonic process. These authors found that this environmentally friendly treatment enhanced anti-inflammation activities two-fold compared to those of conventional extracts processed at high temperature. They proved that ultrasound extraction also showed relatively low cytotoxicity against murine microglial cells and inhibited the production of the inflammatory mediators, NO and PGE_2. The authors also proved that ultrasound extraction effectively suppresses both mRNA expression and the production of proinflammatory cytokines, such as TNF-α, IL-6 and IL-1β, in a concentration-dependent manner. This study also provided useful information for developing functional foods from heat-labile natural resources.

Santos and coworkers [5] developed a low-cost, scalable fermentation protocol to produce a disinfectant from whey (a cheese by-product/dairy waste) with high levels of lactic acid and antimicrobial peptides produced by lactic acid bacteria. They showed that the established fermentation for industrial whey held strong potential as an effective disinfecting agent when applied to lettuce, with better results than 110 ppm chlorine solution. Other advantages were that it did not alter the quality parameters of the shredded loose-leaf lettuce, did not notably affect the color, and panelists were not able to discriminate from chlorine treatments. The authors stated that fermented whey was indeed as effective as chlorine, but also corroborated that their technology of whey fermentation was effective in maintaining the quality of lettuce throughout storage.

Graça and coworkers [6] assessed dairy products as an innovative alternative to enhance the functional and nutritional value of bakery products. Specifically, they studied the addition of yoghurt and cheese curd to wheat bread. These authors found that the yoghurt or cheese additions had a positive impact on the rheology characteristics of the dough. They stated that these enriched breads showed a significant improvement on the nutrition profile, which is important to balance the daily diet in terms of major and trace minerals, and which is critically relevant for health enhancement and maintenance. Moreover, they found good sensorial acceptability for breads with 50 g of yoghurt and 30 g of cheese curd.

Fradinho and coworkers [7] focused on seaweeds as a novel source of important nutritional compounds with interesting biological activities for being processed into added-value products. In this study, two previously developed products obtained by *Laminaria ochroleuca* brown seaweed processing (liquid extract and a purée-like mixture) were processed with *Psyllium* gel to develop functional hydrogels. The optimization of the formulation and the characterization of the *Laminaria–Psyllium* gels in terms of their mechanical features have allowed the proposal of potential food applications. Authors found a beneficial interaction between *Laminaria* and *Psyllium* in terms of the reinforcement of texture and rheological properties, which could provide new healthy gelling formulations with attractive properties to alleviate the growing market demand of eco-novel food matrices.

Funding: This research received no external funding.

Acknowledgments: I.S. and A.R. thank FCT for the financial support of LEAF—UIDP/04129/2020. M.D.T. thanks the Spanish Ministry of Science, Innovation and Universities for their postdoctoral grant (RYC2018-024454-I).

Conflicts of Interest: The authors declare no conflict of interest.

References

1. Faccio, E.; Fovino, L.G.N. Food Neophobia or Distrust of Novelties? Exploring consumers' attitudes toward GMOs, insects and cultured meat. *Appl. Sci.* **2019**, *9*, 4440. [CrossRef]
2. Huang, M.; Zhao, C.; Zou, Y. Increased grain amino acid content in rice with earthworm castings. *Appl. Sci.* **2019**, *9*, 1090. [CrossRef]
3. Garbowska, M.; Pluta, A.; Berthold-Pluta, A. Antihypertensive peptide activity in Dutch-type cheese models prepared with different additional strains of *Lactobacillus* genus bacteria. *Appl. Sci.* **2019**, *9*, 1674. [CrossRef]
4. Choi, W.Y.; Sim, J.-H.; Lee, J.-Y.; Kang, D.H.; Lee, H.Y. Increased anti-inflammatory effects on LPS-induced microglia cells by Spirulina maxima extract from ultrasonic process. *Appl. Sci.* **2019**, *9*, 2144. [CrossRef]
5. Santos, M.I.S.; Fradinho, P.; Martins, S.; Lima, A.I.G.; Ferreira, R.M.S.B.; Pedroso, L.; Ferreira, M.A.S.S.; Sousa, I. A novel way for whey: Cheese whey fermentation produces an effective and environmentally-safe alternative to chlorine. *Appl. Sci.* **2019**, *9*, 2800. [CrossRef]
6. Graça, C.; Raymundo, A.; Sousa, I. Wheat bread with dairy products-technology, nutritional, and sensory properties. *Appl. Sci.* **2019**, *9*, 4101. [CrossRef]
7. Fradinho, P.; Raymundo, A.; Sousa, I.; Domínguez, H.; Torres, M.D. *Psyllium* and *Laminaria* Partnership-An overview of possible food gel applications. *Appl. Sci.* **2019**, *9*, 4356. [CrossRef]

Review

Food Neophobia or Distrust of Novelties? Exploring Consumers' Attitudes toward GMOs, Insects and Cultured Meat

Elena Faccio * and Lucrezia Guiotto Nai Fovino

Department of Philosophy, Sociology, Education and Applied Psychology, University of Padua, Via Venezia 14, 35131 Padua, Italy; lucrezia.guiottonaifovino@gmail.com
* Correspondence: elena.faccio@unipd.it; Tel.: +339-04-9827-7421

Received: 7 September 2019; Accepted: 12 October 2019; Published: 19 October 2019

Abstract: The food industry is constantly challenged to find new ideas to satisfy the increasingly specific consumer demand. However, innovative food products do not always become part of consumption habits or create a market. One of the major sources of resistance to novelty lies in the attitude of the consumer, who in many cases may be suspicious or hostile as a result of specific ideologies, overly attached to tradition, or affected by neophobia. This paper analyzes the construct of food neophobia (the "unwillingness to try new foods") in its phenomenology and its actual power to explain hostility to innovation in the agri-food sector. The limits of the concept, which is not always sufficient to shed light on the many reasons that could underlie the rejection of certain foods, will also be discussed. In addition, we review the recent literature on Europeans' attitude toward novel foods and innovation including Genetically modified organisms (GMOs), cultivated meat and insects as food. This literature reveals a number of paradoxes in consumers' behavior, and in the many complex conditions underpinning the success of innovation in food production. These conditions can only be understood by reconstructing the meanings consumers assign to food, and are often embedded in larger social and political frameworks.

Keywords: food neophobia; food innovation; resistance to novel food; consumer attitudes; GMO; cultured meat; insect consumption

---

1. Introduction

With a global population expected to reach 9 or even 10 billion by the year 2050 and natural resources for food production already scarce in many parts of the world [1,2], humankind is facing a serious challenge: how can we feed everybody? The scientific community has presented various strategies for using current resources to increase the sustainability of food production without using more agricultural land. The main strategies are dietary changes to plant based alternatives, improvements in technology and management and reduction in food waste (e.g., [3,4]). These measures call for changes on the part of both industries/retailers and people involved as customers and as scientists. We investigate several current trends in European thinking about food, seeking to link the various factors involved in consumer acceptance of new dietary elements that are of interest from a sustainability standpoint.

The process of industry innovation is closely linked to the availability of new products on the market. To create consumer demand, new food product launches must combine technological innovation with a series of social and environmental changes, large and small [5]. Indeed, innovation is born from the continuous interaction between the food industry and the institutional and social context in which it operates: it is an opportunity to meet the needs of citizen-consumers, while also responding to emerging social challenges such as environmental sustainability and animal welfare [6–10]. By contrast, research on market trends in food choices shows that there is some resistance to innovations and confirms the stability of decision-making processes that proceed from inertia. Consumers themselves are the biggest obstacle to innovation in the food sector. By definition, an innovative food entails a change from known characteristics, and this tends to clash with habit-bound consumer behavior. This makes many innovations in the agri-food sector incremental rather than radical [7].

As the literature shows, however, it is not always easy to distinguish between cases where consumers' resistance can be circumvented by improving specific product characteristics [11], and cases where resistance is more deeply rooted, as it is linked to specific ideologies about food and is thus not readily overcome. Accordingly, it can be useful to analyze the various facets of food neophobia, or "human unwillingness to consume unfamiliar food" [12], which cuts across the technological challenges now facing the European food industry. As food neophobia can be elicited by a wide range of foods in an increasingly demanding and specialized food market, it can be a major obstacle for producers and consumers alike. Knowing which products are generally associated with neophobia, and which details of the products trigger it is important because it can provide the industry with insight into how to overcome or curb sales resistance. It is equally important to understand whether food neophobia is just an extreme situation, a clinical condition consisting of unjustified rejection of unfamiliar food products, with no logical justification for the adverse emotional reaction of disgust and refusal, or whether it reveals an ingrained attitude that may extend to a wide range of situations.

We will take a closer look at neophobia in order to analyze the construct, understand the instruments whereby it is measured, and explore the situations to which it applies. In fact, neophobia is marked by an underlying "mistrust" of foods that have never been tasted, and is probably linked to a pre-coding of food in a rigid visual schema. Above all, however, it is associated with food's olfactory impact, which jeopardizes its acceptability [10]. This clinically extreme condition mainly affects children and elderly adults, and thus does not seem to involve people who are active purchasers-consumers.

This review of studies on the subject highlights some of the limitations of using the term indiscriminately to explain consumer choices in general. The literature seems to show that feelings of "fear" or "disgust" towards the new in many cases reflect very specific ideological or value choices, which may cause the individual to take up positions in favor of environmentalism, animal welfare or ecology or simply because an individual is clinging to tradition. Thus, we can understand the reasons for avoiding or rejecting a food category only if we identify the consumer's underlying "food philosophy". Moreover, it is only in this way that we can circumvent avoidance and rejection. Confining ourselves to investigating food avoidance behavior could lead us to mistake what is in fact disdain or rebellion for fear. For this reason, our literature review will go into the details of some of these value positions, using other constructs linked to the type of diet chosen to investigate different consumer attitudes to genetically modified organisms (GMOs), cultured meat and edible insects.

2. Processes behind Food Innovation: The Role of Research, Development and Market Orientation in Food Company Innovation and Consumer Demand

Food innovation arises from the interaction of producers, retailers and consumers. Each of these actors brings his own set of requirements and goals, explicit or otherwise, generating highly complex multifactor processes that we will now summarize.

For successful companies, innovation is the key to combining long-term profitability, corporate growth and continuity if it can create stable or growing market demand [10,13]. In a broad sense, the "philosophy of Innovation" includes placing new or improved products and new services on the market, and introducing practices that oblige the company to review its production and organizational system [14]. In addition, it entails the co-construction of a shared lexicon and a system of values that resonates with the consumer's outlook. In fact, some products would have a large market in certain cultural settings, and no market at all in settings that are too far removed from the logic that produced them. If well managed, the set of all these variables enables companies to set themselves apart from the rest of their marketplace [15,16], increase competitiveness and reduce production costs [17].

What distinguishes the companies that are most successful in implementing the philosophy of innovation? Undoubtedly, these companies focus on research, development and market orientation [18]. However, the companies that invest large amounts in research and innovation are not necessary the most innovative [19]. According to Capitanio [7], Research and Development (R&D) also hinges on the quality of human capital. The agri-food sector in particular requires a workforce with different skills and various kinds of human capital—called "inter-functional teams"—in order to innovate effectively [10,20]. Thus, companies' R&D work cannot be assessed merely on the basis of their financial involvement [21]. Market orientation is also vital, viz. "the detection and fulfilment of unfilled needs and wants of potential customers using the skills, resources and competences of the company" [22]. The literature on market orientation argues that manufacturing companies' success is closely linked to being able to rely on the receptiveness of a large market and to the company's skill at satisfying its needs [5,23,24]. Though it is true that the growth of the agri-food sector is not always the result of technological implementation and innovation, its expansion makes it possible to test new products with less risk of failure [25]. For small and medium-sized enterprises, important factors involved in growth and innovation include the characteristics of the entrepreneur and cooperation [26], as well as the market power exercised by retailers [27,28] thanks to networks between retailers, transformers and producers [29]. Through exchanges between retailer and consumer, a steady stream of information regarding the buyers' attitude towards the products placed on the market can be acquired, providing feedback whereby products can be adapted on the basis of consumer satisfaction. Consumer acceptance and willingness to try food innovations involve multiple dimensions, including the values connected to the food choice, the attitude towards specific products [30,31], expectations and economic considerations. Consumer expectations of product quality, sensory characteristics and production processes are crucial for product innovation [32]. Conflicting attitudes towards food innovations that have repercussions on the consumer's decision-making process can arise from the perceived relationship between risks and benefits and comparison with available alternatives [33]. On a food scene crowded with increasingly specific products, the consumer's final decision is often influenced by advertising. Media communication on the subject has always been an important tool for swaying public opinion. In this connection, several scholars have recently drawn attention to the so-called "crisis of authority, trust and responsibility" [34], arising worldwide as a result of the climate fomented by the media, which fuels anxiety about diet, health and food in particular, while at the same time generating a virtual space where anticonformist views can be expressed and strengthened [35–37]. The result is a "dietary cacophony", as Fischler, puts it, or in other words, a continual bombardment of conflicting messages that make it difficult to get our bearings and acquire structured information about what we eat [38]. Oftentimes, the celebrity quick-fix [39] takes on greater resonance in the media than the views of the scientific community. Some studies [40] suggest that scientific communication should avoid an over-polite style that may often be seen as unconvincing. Rather, it may be advisable to take more neutral or even aggressive attitudes, depending on whether the audience has no clear position on the matter, or tends to share the opinion the writer seeks to reinforce. This is a critically important finding, as it alludes to the risk of turning scientific information into a tool for manipulating consumers' emotions. Thus, careful attention should be given to determining the most appropriate ways to inform consumers and promote critical thinking on their part.

3. Food Neophobia

Recent literature [41] has addressed various degrees and types of "aversion to new foods", or food neophobia, which has been defined as the reluctance to eat, or the avoidance of, new or unknown foods. In the following pages, we will reconstruct the origins of the construct and the situations to which it can be applied.

Rozin, the author who first described food neophobia, assumed it has an adaptive and evolutionary function. According to Rozin, human beings are omnivores and, therefore, eat many things. This means that they must use some strategy for avoiding poisonous foods, and preferring foods that will be beneficial to their health and growth [42,43]. In evolutionary terms, this function is fulfilled by neophobias from the moment a child begins to move independently of its parents. Food neophobia thus provides a means of guiding the child towards foods that are already familiar, and rejecting those that are new and might be dangerous. The aversion to bitterness, for example, due to hedonic neurobiological mechanisms present since birth [44], would help the child avoid eating potentially poisonous plants [45], and can persist until adulthood [46]. "Enemy food" may be rejected before tasting, on the basis of vision alone [47]. This has led to the idea of a very rigid visual coding of the food stimulus. If food is recognized as such because it is similar in shape and color to previous favorable food experiences, it is accepted; otherwise it is rejected. A parent's disapproval of food refusal by the child may be associated with the reaction of disgust with food, and that this may make the child even less willing to try new foods [48]. As many as eight to 15 repeated opportunities to sample an item may be needed to learn to accept a previously rejected food. With young children, one opportunity is enough to double the likelihood of consuming new food [49]. As children age, they tend to be less willing to accept new foods [50,51]. The most critical phase seems to be between two and six years [52]. Since repeated exposure can improve willingness to accept new foods, a number of intervention programs have been developed for schools. The "Food Dudes" program, for example, uses rewards, peer-modeling and repeated exposures to fruit and vegetables, two of the food categories that are most often rejected because of neophobia, to the detriment of health. These interventions seem to be effective in reducing food neophobia when deployed as early as possible and in any case not later than the age of nine years and, above all, if they last for at least six months [53,54]. Under these conditions, they seem able to instill a liking for fruit and vegetables [55–58].

There is also another type of behavior, especially among young children, which can sometimes be confused with food neophobia. This is as "picky/fussy" eating [59,60]: the rejection of a large proportion of familiar (as well as novel) foods, resulting in a habitual diet consisting of a particularly small variety of foods. However, as Taylor et al. wrote [61], "there is no single widely accepted definition of picky eating, although most definitions include an element of restricted intake of familiar foods, sometimes with a further degree of food neophobia". In addition, the behavior of super-tasters might overlap or be confused with that of picky/fussy eaters. The fact that many terms are used to describe the same phenomenon (picky, fussy, faddy or choosy eating) has led to the development of several different measures, making it even more difficult to compare data or establish where picky eating ends and food neophobia begins.

3.1. Neophobia in Adults

While food neophobia tends to disappear in adolescence, it can still be found in adults who restrict their diets to a few familiar products and refuse to eat anything else. As a result, they may be subject to nutritional deficiencies or social exclusion. Neophobia in adults appears to be influenced by different socio-demographic variables: urbanization is negatively correlated with neophobia, as is income and schooling [62]. Neophobia mostly affects older people and children, and is less common among young people, especially those who live in cities. It also tends to increase with age; the new generations have become accustomed to a greater variety of foods, both traditional and ethnic [63]. A neophobic component in old age can be due to several factors, including dental problems or gastrointestinal difficulties that can lead the elderly to avoid many foods.

The relationship with gender is not as clear. Some studies suggest that men are more neophobic than women, which could be linked to a wide range of cultural determinants, such as the time spent cooking [63], while other studies find no significant differences between the genders [62].

3.2. The Role of Olfaction in Neophobia

Neophobic children tend to selectively avoid fruit and vegetables [64]. It has been shown [65] that children develop high sensitivity to taste, and bitterness in particular has been negatively correlated with the amount and variety of consumed fruits and vegetables. Researchers believe that even in adults, the sense that is most strongly involved in the development of neophobia is smell [66].

Some studies have shown that people with food neophobia are less able than the general population to perceive odors (whether related to food or not) as pleasant and intense. They also seem to be less willing and interested in tasting new foods and much more uncertain about identifying unknown foods [67,68]. In addition, neophobic eaters generally expect new foods to be less palatable than familiar ones [69].

3.3. Neophobia and Attitude towards New Foods

Can the neophobia construct effectively explain resistance to new foods? Some research data indicates that the relationship between the two is problematic. For example, it has been shown that neophobia does not impact the choice of genetically modified products, or the choice of organic food [70], while it seems to be a negative predictor of willingness to taste non-traditional ethnic foods [71,72] and functional foods, i.e., foods that provide additional health benefits [73,74]. Two other studies found that higher levels of neophobia correlate with less willingness to eat healthy food (vegetables and fruit) in children [75] and in adolescents [76]. The question thus arises as to whether the resistance to accepting new foods is linked to certain food groups and not to newness as such. In other words, since resistance is shown towards fruit and vegetables in particular, it may not impact choices involving new foods. Rather, it may affect food categories that people are already used to eating, such as flour-based or highly processed foods. This contrasts with the findings of the survey conducted by Elorinne [77], which indicated that products of non-animal (i.e., vegetable) origin trigger lower levels of food neophobia than animal-based products. Food neophobia seems to be an extremely complex attitude. Its intensity tends to fluctuate during the life-span and is mediated by several variables [68,78,79]. As for its association with other psychological variables, it correlates with lower scores for sensation seeking [78] and trait anxiety [59], and is negatively correlated with openness to experience [80]. However, little is known about what contributes in the early stages to making a person neophobic, and what factors tend to maintain avoidance over time.

3.4. What Do Neophobia Scores Actually Measure?

Food neophobia is generally investigated using the Food Neophobia Scale (FNS) [81], which has been revised and adapted over the years [63] and translated into various languages. There is also a version for children adapted by Pliner [82]. This instrument investigates the degree of agreement or disagreement with items relating to the consumption of new foods or their avoidance. Specifically, it refers to the willingness to taste new, ethnic or unusual foods. However, it does not investigate the attitude, the emotional reaction and above all the conscious intentions or the reasons for choosing to avoid a food [83]. The FNS, however, is not the only instrument used to investigate unwillingness to try unfamiliar food. In their review, Damsbo-Svendsen [83] presented a qualitative and quantitative comparison of 13 different tools. They recommend that the researcher take great care in choosing the most appropriate instrument: "no instrument is suitable for measuring all aspects of neophobia" (p. 366). Each available instrument explores a different aspect, such as the attitude toward specific food categories (fruit and vegetables) or food situations, or foods produced with new technology. However, for some subcategories of novel food—such as genetically modified and functional foods —the relationship between the consumer's attitude and food neophobia may not be straightforward [67]. Furthermore, less conventional food choices, such as a vegan diet, can also result in high neophobia scores, as Elorinne's survey found [77]; this does not shed light on the reasoning behind the type of choice. The construct thus appears to be poorly defined, as it is very elastic and could encompass food philosophies that are very distant from each other. It would not be unreasonable to hypothesize that it refers more to choices relating to ethnic foods than to willingness to try unfamiliar foods. It should also be borne in mind that the philosophy of "short food supply chains" and locavorism could translate into resistance to new foods. Where such ethical-social considerations are the basic criterion for choosing certain specific food products over others, taking them into account would make it possible to predict consumer behavior more accurately. We also believe that the term "novel food" requires clarification. "Novel" can mean many things: (1) new compared to traditional cuisine, and thus exotic or ethnic food, (2) new compared to what the individual habitually eats, (3) new in the sense of being offered by a different brand, so that the individual tends to eat foods that are similar in their basic ingredients, but produced by other suppliers. Future research should also consider other variables that may play a role in explaining food conduct: it would be important to have a broader conceptualization of the implicit and explicit theories that lead people to engage in certain food behaviors rather than others. This would provide the agri-food industry with guidance regarding the product characteristics that can attract consumers, including those belonging to specific categories such as vegans, etc.

4. Vegetarians' Attitudes to Insect Consumption—Neophobia or Respect for Life?

Insect-based food has a low ecological impact and a high nutritional value, and is thus a potential sustainable alternative for human nutrition [30]. However, consumers' interest in insect-based products is weak [31], as Western food culture generally considers insects disgusting and inappropriate as food [30]. Vegetarians are an under-investigated but interesting population in relation to the consumption of insects. The vegetarian or vegan diets are based on different beliefs about animals that could influence attitudes towards consuming insects in a variety of ways [84,85]. People can fluctuate between different versions of the vegetarian diet. The vegan diet stands out among them, as it is more restrictive than a merely vegetarian diet, since it involves no animal derivatives of any kind [86], and usually entails rather strict ethical positions that are generally very strong in terms of personal identification [87–91]. In a study conducted in Finland, Elorinne et al. [90] examined groups with different diets (vegans, non-vegan vegetarians and omnivores) to compare the consumer's attitude towards insects as food, the influence exerted by social expectations (which the authors call the participants' "subjective norm") on insect consumption, participants' perceived control over their own eating behavior (the three factors underlying Ajzen's "Theory of Planned Behavior (TPB)") [92] and the level of food neophobia. Non-vegan vegetarians and omnivores shared similar values, while vegans differed significantly in almost all investigated constructs. In particular, the three groups interpreted the construct of "responsibility" very differently: vegans consider it very important and translate it into the choice of not eating insect-based food, whereas for omnivores and non-vegan vegetarians, responsibility coincides with environmental sustainability. Vegans also consider eating insects to be morally wrong, given their general tendency to consider meat consumption as more of a moral issue than one of sustainability, which leads them to be disgusted [87]. Vegans are also significantly more neophobic than non-vegan vegetarians and omnivores, a finding which the authors maintain may be partly explained by the fact that vegans' stricter moral attitude towards food of animal origin leads them to rule out eating insects more categorically than non-vegan vegetarians do. Unlike omnivores, none of the vegans mentioned disgust as a reason for refusing to consume insects despite their high neophobia scores, while all cited ethical reasons. This does not necessarily mean that disgust does not figure among the reasons, but it is not among the criteria that respondents considered relevant. As the literature testifies, vegans and non-vegan vegetarians tend to express feelings of disgust about meat consumption [90], but the reasons behind this disgust have not been scrutinized. As a result, a choice of an ethical-moral stamp has been confused with a food "phobia".

Among the general population, women are less likely than men to consume insects [93–95]. In the Finnish survey discussed here, highly educated city-dwelling women accounted for most of the sample, which may have influenced the results. However, as the authors point out, more information is needed about the dietarian identity profiles of consumer groups. Food conduct, as the authors emphasize, is not the outcome of an impromptu choice, simply linked to a particular situation. Food conduct is part of a "dietarian career" [77], i.e., dietary schemas that call for consistency [96] and are developed gradually over time, offering support and reinforcing a position that at first seems only exploratory. We believe that investigating this dietarian career can lay the foundations for arriving at the roots of food choices and understanding the principles they reflect.

5. Synthetic Meat: Does Innovation Taste Better to Those Who Have Never Tasted It?

Sensitivity to the suffering and killing of farm animals is increasing worldwide [87], as is the number of vegetarians. This, however, has not extinguished the desire to eat meat, particularly among higher-income consumers, who nevertheless also declare that they do not want to contribute to animal suffering [97]. From this perspective, cultivated meat is an excellent compromise for protecting animal welfare and allaying the ethical concerns of meat consumers [98]. Cultivated meat, also known as in-vitro, synthetic or "clean" meat, is another example of a novel food. It is produced from animal cells taken from a living animal and then grown in a laboratory environment with a nutrient serum [99–102]. Cultivated meat appears to be a more sustainable alternative to traditional meat: the first studies spoke of its potential to reduce land use by 99%, water use by 96% and energy consumption by up to 45% [103]. More recent research has scaled back these performance estimates, finding that cultivated meat has a smaller land footprint than beef, and lower greenhouse gas emissions than poultry, pork and beef. This, however, comes at the price of a higher energy consumption than that is required for poultry and pork, and which is ultimately comparable to that necessary for beef [104]. The controlled production environment in which cultivated meat would be produced could provide opportunities for health and safety improvements, reducing the risk of diseases [105,106]. Nevertheless, some authors [100,107] point out that large-scale cell culture is never perfectly controlled and that unexpected biological mechanisms, such as the proliferation of cancer cells, may occur in production. That this is in fact a problem for the health of the consumer is still to be demonstrated, but the authors anticipate that it would be a very sensitive topic both for consumers and for legislators tasked with regulating meat cultivation.

Although cultivated meat is unlikely to appear on the market soon, companies are already investigating the profile of potentially interested consumers. Europeans appear to be divided, with at least half deeply suspicious of cultivated meat because it is unnatural [108,109]. It seems that providing information, particularly regarding environmental benefits, is important in order to encourage positive opinions among potential consumers [110]. The lack of familiarity with new technologies has been cited as a cause of distrust, uncertainty and concern about potential long-term negative consequences [110–112]. Both GMOs and cultured meat are categorized in this way, i.e., as technological innovations that arouse feelings of distrust and concern [110]. It would thus be important to reduce the false equivalences created by consumers, for example those that associate cultivated meat with GMOs [111].

Italy provides a good vantage point for investigating attitudes towards cultivated meat, as its food culture centers on traditional and natural foods [113]. In a study conducted by Mancini [114], participants were given information about cultivated meat's positive effects on the environment and showed a generally positive attitude. The best potential consumers were found to be young adults, with a high level of education; if previously informed, they were significantly more interested in buying than the other categories of participants. One of the most intriguing contradictions noted by the researchers is the fact that non-meat eaters had higher expectations about the taste of cultivated meat than meat eaters, even though the latter had expressed a higher intention to buy. Similar findings, with vegans and vegetarians more positive about cultivated meat but less interested in trying it than meat eaters, have also emerged from studies conducted in the United States [115]. The explanation offered for this apparently contradictory behavior is that these categories are not opposed to cultured meat but at the same time are not interested in consuming it. This would be perfectly in line with the choice of a vegan or vegetarian diet, and it thus might be appropriate to direct some strands of research to investigating people's motivations for making certain food choices, which are probably driven by a very strict internal logic even if they seem contradictory at first glance. If this is true, positive perceptions expressed about novel foods should not be simplistically interpreted as an indicator of their potential for commercial success [114].

6. GMOs: Friends or Foes?

GMOs are widely considered to be the future of food. As such, they have been the focus of public debate about their perceived risks, ranging from the reduction of biodiversity to long-term health consequences, such as toxicity or allergies [116,117]. Genetically Modified (GM) foods are derived from plants, animals or microorganisms whose genetic material (DNA) has been artificially modified, e.g., by introducing a gene from other organisms (viruses, bacteria, other plants and animals and even humans). Currently, such modifications are mostly applied in plants to improve their resistance to disease and/or tolerance to herbicides [118]. Because of the controversies both within the scientific community and in public discourse [118–122], many European countries have not yet formally authorized GM crops. According to Boccia [123], Europeans remain for the most part wary of GMOs, with wide regional variations. The highest percentages of opponents of GM foods are in Austria, Norway, Hungary (70%), Cyprus (76%), Italy (77%) and Greece (81%), followed by France and Denmark (65%). The lowest proportions are in Portugal, Ireland, Spain and Finland [124]. These tendencies, however, are not stable over time, and the data is probably heavily influenced by the research design [125–131].

Popek et al. [132] published a study in 2017 on consumer opinions of genetically modified foods conducted in London and Warsaw. The study sought to determine whether there are cultural differences between the two countries, whose history in terms of GMO acceptance is quite dissimilar. Poland, which banned GMO crops in 2015, also restricts trade in genetically modified organisms. England is among the leaders in biotechnology, and GMOs boast the full support of the government, which sees them as an important economic resource. Contrary to expectations, the study's results were very homogeneous, suggesting that since all European consumers have access to global information sources, their attitudes towards certain phenomenon tend to converge. In both countries, city dwellers were similar in their attitudes and were more favorable to GMOs than their warier rural compatriots. This might merit further investigation. Most respondents agreed that longer shelf-life and resistance to extreme weather conditions are the main advantages of GMOs, while the most feared disadvantages were the unpredictable consequences of genetic modification, the production of species-specific toxins and food allergenicity. Other studies [133] found different concerns, including the fear of carcinogenic effects, environmental damage and the disappearance of natural products from the market. Overall, as many as 27.69% of respondents surveyed by Popek et al. had a negative attitude towards GM foods, while only 19.83% believed that GMOs would bring tangible benefits. The other respondents expressed no definite opinion.

The literature presents conflicting evidence about the role of information in changing attitudes towards GMOs. According to Scholderer and Frewer [53], none of the information strategies implemented in different European countries has succeeded in changing attitudes; indeed, it seems that they have negatively influenced the choice of products. Attitudes towards GMOs do not improve even when the mechanisms underlying the genetic modification are understood. It is clear that this is a particularly complex topic, and that consumer attitudes are probably also constructed on the basis of dimensions that research has not considered, which could involve social dynamics and values (i.e., in rural areas) that cannot be changed merely by providing cognitively relevant information. Yuan provides some innovative thoughts about using communication styles tailored to the audience to maximize the effectiveness of a positive message about GMOs. Focusing on the main trends of thought that are reinforced at the cultural and community level could provide more insight into the meanings underlying this generalized hostility. If this is indeed where the answer lies, measures intended to change individuals' opinions simply by providing information could be much less effective than hoped, as meanings are also socially constructed [134].

7. Discussion

It is clear from the studies analyzed here that the relationship between neophobia and technological innovation in the agri-food industry is much more complex and nuanced than it might seem. Using the concept of neophobia outside of its original clinical context to explain the motivation for consumer choices has many limitations. Consumer choices stem from systems of values that are rich in moral implications, often linked to sociopolitical and ecological-environmental considerations. Such judgments go far beyond what is thought to be the original adaptive value of neophobia. They cannot be reduced to the "picky eating" typical of children, given that food is also closely linked to identity and has a profound social and cultural meaning. The items in the Food Neophobia Scale [63,81,82] investigate the consumption or avoidance of certain foods, referring in particular to new, ethnic or unusual foods. They do not attempt to tap aspects relating to attitude or emotional reaction. This makes it impossible to gain even a superficial grasp of the rationale behind the choice of avoidance, which could hinge on a wide range of reasons. Unconventional diets, such as the vegan diet, could result in high neophobia scores but are not situations to which the term neophobia applies, since they are choices based on ethical values and not simply on disgust toward specific foods. Moreover, what constitutes "novel foods" is poorly specified, as the term novel is used indiscriminately in the literature to denote food that is new compared to traditional cuisine and thus exotic or ethnic; new compared to what the individual habitually eats; or new in the sense of being offered by a different brand, similar in basic ingredients, but produced by other suppliers. We then examined some recent data on potential "novel foods", presenting several points that may be of interest in scrutinizing the meanings and motivations behind certain food choices.

Insects, which could provide an alternative source of nutritious proteins with a lower environmental cost than traditional animal protein sources, are a novelty to the European food scene. They are also particularly likely to elicit neophobic feelings, as they are considered exotic, "disgusting" and foreign to European food culture. In studies of the attitudes towards insect consumption of people with different eating styles (omnivores, vegans and non-vegan vegetarians), vegans were found to have the highest neophobia scores, not because they express disgust with insects, but out of consistency with their ethical objections to eating animals or animal derivatives. Much more favorable attitudes were held by non-vegan vegetarians, who are more concerned with environmental sustainability than with animal rights, and seem to feel that insects are not "proper" animals and can thus be eaten.

Lastly, consumers are divided between acceptance and rejection of synthetic meat. Though still utopian, synthetic meat is a solution that would address some of the most pressing problems associated with meat eating by eliminating animal exploitation and reducing the amount of energy and land needed to produce meat. Though cultivated meat's environmental benefits could be the key to swaying potential consumers' attitudes in its favor, this clashes with a general distrust of its "unnaturalness" and the potential consequences of the new technologies on health. It seems that the consumers with the greatest interest in buying are young, well-educated and knowledgeable about cultivated meat. In an apparent paradox, it has been found that vegetarians who were not interested in sampling cultivated meat had higher expectations about its taste than meat eaters who were in fact interested in buying. Future research could profitably address the motivations behind food choices, which prevent a positive perception from being transformed into an intention to buy, complicating the scene surrounding this novel food.

GMOs, long at the center of a series of controversies in the public discourse and in its legislative repercussions, still inspire conflicting opinions, some of which are uncompromisingly negative. A significant proportion of consumers in European capitals state that they are worried about GMOs' impact on health, biodiversity and the environment, though a majority has no firm opinion on the matter. This suggests that the information strategies implemented in previous decades have not had appreciable effects. It is doubtful that providing information can change attitudes to GMOs, and it may even be counterproductive. This may be because scholars often over-compartmentalize the motivations that then become attitudes: rather than being considered as part of an organic complex that also involves values and social dimensions, motivations are reduced to single affective or cognitive components. For example, it seems that where one lives could be a major predictor of GMO acceptance. Addressing this topic in future research would provide information whereby strategic marketing can be tailored to specific targets. Conceivably, this could improve the efficiency of a persuasion process aimed at increasing the acceptance of GMO technology.

8. Conclusions

This review of the recent situation in several European countries has explored several elements that are crucial for the development and innovation in the agri-food sector, both from the standpoint of the industry and from the consumer's perspective. In the latter connection, though consumer acceptance or rejection of a new food product hinges on a complex range of variables, three conclusions can be drawn that can assist companies that intend to introduce innovative practices or new products on the market:

(1) Resistance or refusal of new foods is rarely motivated by neophobia.
(2) More often, the rejection of a food choice should not be interpreted as an aversion to the specific product, or as selective and unmotivated avoidance of that product, but as an operational extension of a vision of the product filtered through the system of values embraced by the individual [134].
(3) In contemporary society, when many food alternatives are available, individuals discover and develop their own food identity, which is an important part of their self-identification and values (i.e., whether they see themselves as health-conscious, environmentalists, defenders of animal rights or traditional omnivores). This generates a stable identity profile that tends to be consistent and confirmed over time, in a dietarian career that progresses through successive steps.

Thus, we believe that future research should address the attitude towards new food products starting from an understanding of the food identity profile of the members of the population of interest. Only in this way will it be possible to tap the psychological variables linked to the system of values that drive food choices. While these values manifest themselves in the conduct of the individual, they are always in resonance with a food community, anchored to a series of typifications of food choices that individuals then tend to take as their own. A better understanding of these values would make it possible to determine whether innovations are consistent or compatible with the career profile to which the person tends to correspond. To be effective, information and marketing campaigns should be tailored to specific eater identity profile groups. An underinvestigated strategy for promoting a positive attitude to food innovation consists of investing in food ingredients or technologies that can be viewed favorably by specific food subgroups on the basis of the value system that their "food philosophy" is committed to safeguarding.

As this review has shown, today's food choices are highly fragmented as a result of individuals' differing ideological attitudes. Consequently, expecting to be able to introduce new foods that can be interesting or desirable across all consumer families seems outdated and unrealistic. Choosing the dietary group to be targeted by the innovation as the first step would seem to be a far more promising strategy.

Author Contributions: Conceptualization, review of the literature, investigation, writing—original draft preparation; writing—review and editing, E.F. and L.G.N.F.; supervision: E.F.

Funding: This research received no external funding.

Conflicts of Interest: The authors declare no conflict of interest.

References

1. Gerland, P.; Raftery, A.E.; Ševˇcíková, H.; Li, N.; Gu, D.; Spoorenberg, T.; Alkema, L.; Fosdick, B.K.; Chunn, J.; Lalic, N.; et al. World population stabilization unlikely this century. *Science* **2014**, *346*, 234–237. [CrossRef]

2.   United Nations. *World Population Prospects: The 2017 Revision, Key Findings and Advance Tables*; ESA/P/WP/248; United Nations Department of Economic and Social Affairs/Population Division: New York, NY, USA, 2017.

3.   Kummu, M.; Fader, M.; Gerten, D.; Guillaume, J.H.; Jalava, M.; Jägermeyr, J.; Pfister, S.; Porkka, M.; Siebert, S.; Varis, O. Bringing it all together: Linking measures to secure nations' food supply. *Curr. Opin. Environ. Sustain.* **2017**, *29*, 98–117. [CrossRef]

4.   Springmann, M.; Clark, M.; Mason-D'Croz, D.; Wiebe, K.; Bodirsky, B.L.; Lassaletta, L.; de Vries, W.; Vermeulen, S.J.; Herrero, M.; Carlson, K.M.; et al. Options for keeping the food system within environmental limits. *Nature* **2018**, *562*, 519–525. [CrossRef] [PubMed]

5.   Earle, M.D. Innovation in the food industry. *Trends Food Sci. Technol.* **1997**, *8*, 166–175. [CrossRef]

6.   Mancini, P.; Marotta, G.; Nazzaro, C.; Simonetti, B. Consumer behaviour, obesity and social costs: The case of Italy. *Int. J. Bus. Soc.* **2015**, *16*, 295–324. [CrossRef]

7.   Capitanio, F.; Coppola, A.; Pascucci, S. Product and process innovation in the Italian food industry. *Agribusiness* **2012**, *26*, 503–518. [CrossRef]

8.   Marotta, G.; Nazzaro, C. Responsabilità sociale e creazione di valore nell'impresa agroalimentare: Nuove frontiere di ricerca. *Econ. Agro-Aliment.-Food Econ.* **2012**, *1*, 13–54.

9.   Nazzaro, C.; Lerro, M.; Marotta, G. Assessing parental traits affecting children's food habits: An analysis of the determinants of responsible consumption. *Agric. Food Econ.* **2018**, *6*, 23–37. [CrossRef]

10.   Roucan-Kane, M.; Gray, A.W.; Boehlje, M.D. Approaches for selecting product innovation projects in US food and agribusiness companies. *Int. Food Agribus. Manag. Rev.* **2011**, *14*, 51–68.

11.   Nazzaro, C.; Lerro, M.; Stanco, M.; Marotta, G. Do consumers like food product innovation? An analysis of willingness to pay for innovative food attributes. *Br. Food J.* **2019**, *121*, 1413–1427. [CrossRef]

12.   La Barbera, F.; Verneau, F.; Amato, M.; Grunert, K. Understanding Westerners' disgust for the eating of insects: The role of food neophobia and implicit associations. *Food Qual. Prefer.* **2018**, *64*, 120–125. [CrossRef]

13.   Grunert, K.G.; van Trijp, H.C.M. Consumer-oriented new product development. *Encycl. Agric. Food Syst.* **2014**, *2*, 375–386.

14.   Damanpour, F. Combinative effects of innovation types and organizational performance: A longitudinal study of service organizations. *J. Manag. Stud.* **2009**, *46*, 650–675. [CrossRef]

15.   Baregheh, A.; Rowley, J.; Sambrook, S.; Davies, D. Innovation in food sector SMEs. *J. Small Bus. Enterp. Dev.* **2012**, *19*, 300–321. [CrossRef]

16.   Karantininis, K.; Sauer, J.; Furtan, W.H. Innovation and integration in the agri-food industry. *Food Policy* **2010**, *35*, 112–120. [CrossRef]

17.   Marotta, G.; Nazzaro, C.; Stanco, M. How the social responsibility creates value: Models of innovation in Italian pasta industry. *Int. J. Glob. Small Bus.* **2017**, *9*, 144–167. [CrossRef]

18.   Grunert, K.G.; Harmsen, H.; Meulenberg, M.; Kuiper, E.; Ottowitz, T.; Declerck, F.; Traill, B.; Göransson, G. A framework for analysing innovation in the food sector. In *Product and Process Innovation in the Food Sector*; Traill, B., Grunert, K.G., Eds.; Blackie Academic: London, UK, 1997; pp. 1–37.

19.   Bhattacharya, M.; Bloch, H. Determinants of innovation. *Small Bus. Econ.* **2004**, *22*, 155–162. [CrossRef]

20.   Cooper, R.G.; Edgett, S.J.; Kleinschmidt, E.J. Benchmarking Best NPD practices—II. *Res. Technol. Manag.* **2004**, *47*, 50–59. [CrossRef]

21.   Rama, R. Empirical study on sources of innovation in international food and beverage industry. *Agribusiness* **1996**, *12*, 123–134. [CrossRef]

22.   Grunert, K.G.; Hartvig Larsen, H.L.; Madsen, T.K.; Baadsgaard, A. *Market Orientation in Food and Agriculture*; Kluwer Academic: Boston, MA, USA, 1996.

23.   Borch, O.J.; Forsman, S. The competitive tools and capabilities of micro firms in the Nordic food sector: A comparative study. In *the Food Sector in Transition: Nordic Research*; Norwegian Agricultural Economics Research Institute: Oslo, Norway, 2000.

24.   Steward-Knox, B.; Mitchell, P. What separates the winners from the losers in new product development. *Trends Food Sci. Technol.* **2003**, *14*, 58–64. [CrossRef]

25.   Triguero, A.; Córcoles, D.; Cuerva, C.M. Differences in innovation between food and manufacturing firms: An analysis of persistence. *Agribusiness* **2013**, *29*, 273–292. [CrossRef]

26.   Avermaete, T.; Viaene, J.; Morgan, E.J.; Pitts, E.; Crawford, N.; Mahon, D. Determinants of product and process innovation in small food manufacturing firms. *Trends Food Sci. Technol.* **2004**, *15*, 474–483. [CrossRef]

27. Dobson, P.; Clarke, R.; Davies, S.; Waterson, M. Buyer power and its impact on competition in the food retail distribution sector of the European Union. *J. Ind. Compet. Trade* **2001**, *1*, 274–281. [CrossRef]

28. Weiss, H.R.; Wittkopp, A. Retailer concentration and product innovation in food manufacturing. *Eur. Rev. Agric. Econ.* **2005**, *32*, 219–244. [CrossRef]

29. Cox, H.; Mowatt, S.; Prevezer, M. New product development and product supply within a network setting: The chilled ready-meal industry in the UK. *Ind. Innov.* **2003**, *10*, 197–217. [CrossRef]

30. Fischer, C.B.; Steenbekkers, L.P.A.B. All insects are equal, but some insects are more equal than others. *Br. Food J.* **2018**, *120*, 852–863. [CrossRef]

31. Hartmann, C.; Siegrist, M. Consumer perception and behaviour regarding sustainable protein consumption: A systematic review. *Trends Food Sci. Technol.* **2017**, *61*, 11. [CrossRef]

32. Grunert, K.G. Consumer behaviour with regard to food innovations: Quality perception and decision-making. In *Innovations in Agri-Food Systems—Product Quality and Consumer Acceptance*; Jongen, W.M.F., Meulenberg, M.T.G., Eds.; Wageningen Academic Publishers: Wageningen, The Netherlands, 2005; pp. 57–82.

33. Dolgopolova, I.; Teuber, R.; Bruschi, V. Consumers' perceptions of functional foods: Trust and food neophobia in a cross-cultural context. *Int. J. Consum. Stud.* **2015**, *39*, 708–715. [CrossRef]

34. Belasco, W. Foreword. In *Food Words: Essays in Culinary Culture*; Jackson, P., Ed.; Routledge: London, UK, 2015; pp. 864–865. [CrossRef]

35. Jackson, P.; Watson, M.; Piper, N. Locating Anxiety in the Social: The Cultural Mediation of Food Fears. *Eur. J. Cult. Stud.* **2013**, *16*, 24–42. [CrossRef]

36. Levenstein, H. *Fear of Food: A History of Why We Worry about What We Eat*; University of Chicago Press: London, UK, 2012.

37. Rousseau, S. *Food Media: Celebrity Chefs and the Politics of Everyday Interference*; Berg: Oxford, UK, 2012.

38. Levenstein, H. *Paradox of Plenty: A Social History of Eating in Modern America*; University of California Press: Los Angeles, CA, USA, 2003.

39. Rousseau, S. The celebrity quick-fix: When good food meets bad science. *Food Cult. Soc.* **2015**, *18*, 265–287. [CrossRef]

40. Yuan, S.; Ma, W.; Besley, J.C. Should Scientists Talk About GMOs Nicely? Exploring the Effects of Communication Styles, Source Expertise, and Preexisting Attitude. *Sci. Commun.* **2009**, *41*, 267–290. [CrossRef]

41. Kallas, Z.; Vitale, M.; Gil, J.M. Health innovation in patty products. the role of food neophobia in consumers' non-hypothetical willingness to pay, purchase intention and hedonic evaluation. *Nutrients* **2019**, *11*, 444. [CrossRef] [PubMed]

42. Rozin, E.; Rozin, P. Culinary themes and variations. *Nat. Hist.* **1981**, *90*, 6–14.

43. Rozin, P.; Vollmecke, T.A. Food likes and dislikes. *Annu. Rev. Nutr.* **1986**, *6*, 433–456. [CrossRef]

44. Steiner, J.E. Human facial expressions in response to taste and smell and stimulation. *Adv. Child. Dev. Behav.* **1979**, *13*, 257–295.

45. Glander, K.E. The impact of plant secondary compounds on primate feeding behavior. *Yearb. Phys. Anthropol.* **1982**, *25*, 1–18. [CrossRef]

46. Stein, L.J.; Nagai, H.; Nakagawa, M.; Beauchamp, G.K. Effects of repeated exposure and health-related information on hedonic evaluation and acceptance of a bitter beverage. *Appetite* **2003**, *40*, 119–129. [CrossRef]

47. Harris, G. Introducing the infant's first solid food. *Br. Food J.* **1993**, *95*, 7–10. [CrossRef]

48. Burgess, T.D.G.; Sales, S.M. Attitudinal effects of "mere exposure": A reevaluation. *J. Exp. Soc. Psychol.* **1971**, *7*, 461–472. [CrossRef]

49. Birch, L.L.; Gunder, L.; Grimm-Thomas, K.; Laing, D.G. Infants' consumption of a new food enhances acceptance of similar foods. *Appetite* **1998**, *30*, 283–295. [CrossRef]

50. Cooke, L.J.; Carnell, S.; Wardle, J. Food neophobia and mealtime food consumption in 4–5 year old children. *Int. J. Behav. Nutr. Phys. Act.* **2006**, *3*, 14–19. [CrossRef] [PubMed]

51. Wardle, J.; Cooke, L.J.; Gibson, E.L.; Sapochnik, M.; Sheiham, A.; Lawson, M. Increasing children's acceptance of vegetables: A randomised trial of parent-led exposure. *Appetite* **2003**, *40*, 155–162. [CrossRef]

52. Cooke, L.; Wardle, J.; Gibson, E.L. Relationship between parental report of food neophobia and everyday food consumption in 2-6-year-old children. *Appetite* **2003**, *41*, 205–206. [CrossRef]

53. Scholderer, J.; Frewer, L.J. The biotechnology communication paradox: Experimental evidence and the need for a new strategy. *J. Consum. Policy* **2003**, *26*, 125–127. [CrossRef]

54. Kronberger, N. Moralities We Live by: Moral Focusing in the Context of Technological Change. In *Meaning in Action*; Sugiman, T., Gergen, K.J., Wagner, W., Yamada, Y., Eds.; Springer: Tokyo, Japan, 2008.

55. Laureati, M.; Bergamaschi, V.; Pagliarini, E. School-based intervention with children. Peer-modeling, reward and repeated exposure reduce food neophobia and increase liking of fruits and vegetables. *Appetite* **2014**, *83*, 26–32. [CrossRef]

56. Faccio, E.; Costa, N.; Losasso, C.; Cappa, V.; Mantovani, C.; Cibin, V.; Andrighetto, I.; Ricci, A. What programs work to promote health for children? Exploring beliefs on microorganisms and on food safety control behavior in primary schools. *Food Control.* **2013**, *33*, 320–329. [CrossRef]

57. Faccio, E.; Costa, N.; Losasso, C.; Barrucci, F.; Mantovani, C.; Cibin, V.; Andrighetto, I.; Ricci, A. Drawing instead of answering to evaluate the effectiveness of food safety programmes in primary school. *Health Educ. J.* **2017**, *76*, 15–28. [CrossRef]

58. Losasso, C.; Cappa, V.; Cibin, V.; Mantovani, C.; Costa, N.; Faccio, E.; Andrighetto, I.; Ricci, A. Food safety and hygiene lessons in the primary school: Implications for risk-reduction behaviors. *Foodborne Pathog. Dis.* **2014**, *11*, 68–74. [CrossRef]

59. Dovey, T.M.; Staples, P.A.; Leigh Gibson, E.; Halford, J.C.G. Food neophobia and 'picky/fussy' eating in children: A review. *Appetite* **2008**, *50*, 181–193. [CrossRef]

60. Lafraire, J.; Camille Rioux, C.; Giboreau, A.; Picard, D. Food rejections in children: Cognitive and social/environmental factors involved in food neophobia and picky/fussy eating behavior. *Appetite* **2016**, *96*, 347–357. [CrossRef]

61. Taylor, C.M.; Wernimont, S.M.; Northstone, K.; Emmett, P.M. Picky/fussy eating in children: Review of definitions, assessment, prevalence and dietary intakes. *Appetite* **2015**, *95*, 349–359. [CrossRef] [PubMed]

62. Meiselman, H.L.; King, S.C.; Gillette, M. The demographics of neophobia in a large commercial US sample. *Food Qual. Prefer.* **2010**, *21*, 893–897. [CrossRef]

63. Siegrist, M.; Hartmann, C.; Keller, C. Antecedents of food neophobia and its association with eating behavior and food choices. *Food Qual. Prefer.* **2013**, *30*, 293–298. [CrossRef]

64. Wardle, J.; Cooke, L. Genetic and environmental determinants of children's food preferences. *Br. J. Nutr.* **2008**, *99*, 15–21. [CrossRef] [PubMed]

65. Coulthard, H.; Blissett, J. Fruit and vegetable consumption in children and their mothers. Moderating effects of child sensory sensitivity. *Appetite* **2009**, *52*, 410–415. [CrossRef] [PubMed]

66. Demattè, M.L.; Endrizzi, I.; Gasperi, F. Food neophobia and its relation with olfaction. *Front. Psychol.* **2014**, *5*, 127. [CrossRef]

67. Tuorila, H.M.; Mustonen, S. Reluctant trying of an unfamiliar food induces negative affection for the food. *Appetite* **2010**, *54*, 418–421. [CrossRef]

68. Frank, R.A.; Raudenbush, B. Individual differences in approach to novelty: The case of human food neophobia. In *Viewing Psychology as a Whole: The Integrative Science of William*; Dember, N., Hoffman, R.R., Sherrick, M.F., Warm, J.S., Eds.; American Psychological Association: Washington, DC, USA, 1998; pp. 227–245.

69. Pliner, P.; Pelchat, M.L.; Grabski, M. Reduction of neophobia in humans by exposure to novel foods. *Appetite* **1993**, *20*, 111–123. [CrossRef]

70. Backstrom, A.; Pirttila-Backman, A.M.; Tuorila, H. Willingness to try new foods as predicted by social representations and attitude and trait scales. *Appetite* **2004**, *43*, 75–83. [CrossRef]

71. Choe, J.Y.; Cho, M.S. Food neophobia and willingness to try non-traditional foods for Koreans. *Food Qual. Prefer.* **2011**, *22*, 671–677. [CrossRef]

72. D'Antuono, L.F.; Bignami, C. Perception of typical Ukrainian foods among an Italian population. *Food Qual. Prefer.* **2012**, *25*, 1–8. [CrossRef]

73. Siegrist, M.; Stampfli, N.; Kastenholz, H. Consumers' willingness to buy functional foods. The influence of carrier, benefit and trust. *Appetite* **2008**, *51*, 526–529. [CrossRef] [PubMed]

74. Urala, N.; Lahteenmaki, L. Consumers' changing attitudes towards functional foods. *Food Qual. Prefer.* **2007**, *18*, 1–12. [CrossRef]

75. Falciglia, G.A.; Couch, S.C.; Gribble, L.S.; Pabst, S.M.; Frank, R. Food neophobia in childhood affects dietary variety. *J. Am. Diet. Assoc.* **2000**, *100*, 1474. [CrossRef]

76. Mustonen, S.; Oerlemans, P.; Tuorila, H. Familiarity with and affective responses to foods in 8–11-year-old children. The role of food neophobia and parental education. *Appetite* **2012**, *58*, 777–780. [CrossRef]

77. Elorinne, A.; Niva, M.; Vartiainen, O.; Väisänen, P. Insect consumption attitudes among vegans, non-vegan vegetarians, and omnivores. *Nutrients* **2019**, *11*, 292. [CrossRef]

78. Otis, L.P. Factors influencing the willingness to taste unusual foods. *Psychol. Rep.* **1984**, *54*, 739–745. [CrossRef]

79. Howard, A.J.; Mallan, K.M.; Byrne, R.; Magarey, A.; Daniels, L.A. Toddlers' food preferences. The impact of novel food exposure, maternal preferences and food neophobia. *Appetite* **2012**, *59*, 818–825. [CrossRef]

80. Knaapila, A.; Silventoinen, K.; Broms, U.; Rose, R.J.; Perola, M.; Kaprio, J.; Tuorila, H.M. Food neophobia in young adults: Genetic architecture and relation to personality, pleasantness and use frequency of foods, and Body Mass Index. A twin study. *Behav. Genet.* **2011**, *41*, 512–521. [CrossRef]

81. Pliner, P.; Hobden, K. Development of a scale to measure the trait of food neophobia in humans. *Appetite* **1992**, *19*, 105–120. [CrossRef]

82. Pliner, P. Development of measures of food neophobia in children. *Appetite* **1994**, *23*, 147–163. [CrossRef] [PubMed]

83. Damsbo-Svendsen, M.; Bom Frøst, M.; Olsen, A. A review of instruments developed to measure food neophobia. *Appetite* **2017**, *113*, 358–367. [CrossRef] [PubMed]

84. Fischer, B. Bugging the strict vegan. *J. Agric. Environ. Ethics* **2016**, *29*, 255–263. [CrossRef]

85. Tuorila, H.; Lahteenmaki, L.; Pohjalainen, L.; Lotti, L. Food neophobia among the Finns and related responses to familiar and unfamiliar foods. *Food Qual. Prefer.* **2001**, *12*, 29–37. [CrossRef]

86. House, J. Are insects animals? The ethical position of insects in Dutch vegetarian diets. In *Ethical Vegetarianism and Veganism*; Linzey, A., Linzey, C., Eds.; Routledge: London, UK, 2019.

87. Melina, V.; Craig, W.; Levin, S. Position of the academy of nutrition and dietetics: Vegetarian diets. *J. Acad. Nutr. Diet.* **2016**, *116*, 1970–1980. [CrossRef]

88. Ruby, M.B. Vegetarianism. A blossoming field of study. *Appetite* **2012**, *58*, 141–150. [CrossRef]

89. Fox, N.; Ward, K. Health, ethics and environment: A qualitative study of vegetarian motivations. *Appetite* **2008**, *50*, 422–429. [CrossRef]

90. Radnitz, C.; Beezhold, B.; DiMatteo, J. Investigation of lifestyle choices of individuals following a vegan diet for health and ethical reasons. *Appetite* **2015**, *90*, 31–36. [CrossRef]

91. Elorinne, A.; Kantola, M.; Voutilainen, S.; Laakso, J. Veganism as a choice: Experiences and food strategies in transitioning to a vegan diet. In *Food Futures: Ethics, Science and Culture*; Wageningen Academic Publisher: Wageningen, The Netherlands, 2016.

92. Rosenfeld, D.L. A comparison of dietarian identity profiles between vegetarians and vegans. *Food Qual. Prefer.* **2019**, *72*, 40–44. [CrossRef]

93. Ajzen, I. The theory of planned behavior. *Organ. Behav. Hum. Decis. Process.* **1991**, *50*, 179–211. [CrossRef]

94. Tan, H.S.G.; van den Berg, E.; Stieger, M. The influence of product preparation familiarity and individual traits on the consumer acceptance of insects as food. *Food Qual. Prefer.* **2016**, *48*, 222–231. [CrossRef]

95. Cicatiello, C.; De Rosa, B.; Franco, S.; Lacetera, N. Consumer approach to insects as food: Barriers and potential for consumption in Italy. *Br. Food J.* **2016**, *118*, 2271–2286. [CrossRef]

96. Verbeke, W. Profiling consumers who are ready to adopt insects as a meat substitute in a Western society. *Food Qual. Prefer.* **2015**, *39*, 147–155. [CrossRef]

97. Rosenfeld, D.L.; Burrow, A.L. Development and validation of the dietarian identity questionnaire: Assessing self-perceptions of animal-product consumption. *Appetite* **2018**, *127*, 182–194. [CrossRef]

98. Hopkins, P.D.; Dacey, A. Vegetarian meat: Could technology save animals and satisfy meat eaters? *J. Agric. Environ. Ethics* **2008**, *21*, 579–596. [CrossRef]

99. Bartholet, J. Inside the meat lab. *Sci. Am.* **2011**, *304*, 64–69. [CrossRef]

100. Edelman, P.D.; McFarland, D.C.; Mironov, V.A.; Matheny, J.G. Commentary: In vitro-cultured meat production. *Tissue Eng.* **2005**, *11*, 659–662. [CrossRef]

101. Bhat, Z.F.; Fayaz, H. Prospectus of cultured meat—Advancing meat alternatives. *J. Food Sci. Technol.* **2011**, *48*, 125–140. [CrossRef]

102. Post, M.J. Cultured meat from stem cells: Challenges and prospects. *Meat Sci.* **2012**, *92*, 297–301. [CrossRef]

103. Moritz, M.S.; Verbruggen, S.E.; Post, M.J. Alternatives for large-scale production of cultured beef: A review. *J. Integr. Agric.* **2015**, *14*, 208–216. [CrossRef]

104. Tuomisto, H.L.; Teixeira de Mattos, M.J. Environmental Impacts of Cultured Meat Production. *Environ. Sci. Technol.* **2011**, *45*, 6117–6123. [CrossRef] [PubMed]

105. Mattick, C.S.; Landis, A.E.; Allenby, B.R.; Genovese, N.J. Anticipatory life cycle analysis of in vitro biomass cultivation for cultured meat production in the United States. *Environ. Sci. Technol.* **2015**, *49*, 11941–11949. [CrossRef] [PubMed]

106. Datar, I.; Betti, M. Possibilities for an in vitro meat production system. *Innov. Food Sci. Emerg. Technol.* **2010**, *11*, 13–22. [CrossRef]

107. Hocquette, J.F.; Mainsant, P.; Daudin, J.D.; Cassar Malek, I.; Rémond, D.; Doreau, M.; Sans, P.; Bauchart, D.; Agabriel, J.; Verbecke, W.; et al. Will meat be produced in vitro in the future? *INRA Prod. Anim.* **2013**, *26*, 363–374.

108. Hocquette, J.F. Is in vitro meat the solution for the future? *Meat Sci.* **2016**, *120*, 167–176. [CrossRef] [PubMed]

109. Eurobarometer, S. Social values, science and technology. *Eurobarom. Spec. Rep.* **2005**, 225.

110. Verbeke, W.; Marcu, A.; Rutsaert, P.; Gaspar, R.; Seibt, B.; Fletcher, D.; Barnett, J. 'Would you eat cultured meat?': Consumers' reactions and attitude formation in Belgium, Portugal and the United Kigdom. *Meat Sci.* **2015**, *102*, 49–58. [CrossRef]

111. Verbeke, W.; Sans, P.; Van Loo, E. Challenges and prospects for consumer acceptance of cultured meat. *J. Integr. Agric.* **2015**, *14*, 285–294. [CrossRef]

112. Marcu, A.; Gaspar, R.; Rutsaert, P.; Seibt, B.; Fletcher, D.; Verbeke, W.; Barnett, J. Analogies, metaphors, and wondering about the future: Lay sense-making around synthetic meat. *Public Underst. Sci.* **2015**, *24*, 547–562. [CrossRef]

113. Siegrist, M.; Sütterlin, B. Importance of perceived naturalness for acceptance of food additives and cultured meat. *Appetite* **2017**, *113*, 320–326. [CrossRef]

114. Lazzaroni, C.; Iacurto, M.; Vincenti, F.; Biagini, D. Consumer Attitudes to Food Quality Products of Animal Origin in Italy. In *Consumer Attitudes to Food Quality Products. Emphasis on Southern Europe*; Klopčič, M., Kuipers, A., Hocquette, J.F., Eds.; Wageningen Academic Publishers: Wageningen, The Netherlands, 2013; pp. 83–96.

115. Mancini, M.C.; Antonioli, F. Exploring consumers' attitude towards cultured meat in Italy. *Meat Sci.* **2019**, *150*, 101–110. [CrossRef] [PubMed]

116. Wilks, M.; Phillips, C.J. Attitudes to in vitro meat: A survey of potential consumers in the United States. *PLoS ONE* **2017**, *12*, e0171904. [CrossRef] [PubMed]

117. Bongoni, R. East versus West: Acceptance of GM foods by European and Asian Consumers. *Nutr. Food Sci.* **2016**, *46*, 628–636. [CrossRef]

118. Davis, L.A. Genetically Engineered Crops. *Engineering* **2016**, *2*, 268–269. [CrossRef]

119. Domingo, J.L. Safety assessment of GM plants: An updated review of the scientific literature. *Food Chem. Toxicol.* **2016**, *95*, 12–18. [CrossRef]

120. López, O.A.M.; Pérez, E.F.; Fuentes, E.E.S.; Luna-Espinoza, I.; Cuevas, F.A. Perceptions and attitudes of the Mexican urban population towards genetically modified organisms. *Br. Food J.* **2016**, *118*, 2873–2892. [CrossRef]

121. Sarno, V.; Manzo, R.M. Italian companies' attitude towards GM crops. *Nutr. Food Sci.* **2016**, *46*, 685–694. [CrossRef]

122. Grunert, K.G.; Bredahl, L.; Scholderer, J. Four questions on European consumers' attitudes toward the use of genetic modification in food production. *Innov. Food Sci. Emerg. Technol.* **2003**, *4*, 435–445. [CrossRef]

123. Boccia, F. Consumer perception: An analysis on second generation genetically modified foods. *Nutr. Food Sci.* **2016**, *46*, 637–646. [CrossRef]

124. Vlontzos, G.; Duquenne, M.N. To eat or not to eat? The case of genetically modified (GM) food. *Nutr. Food Sci.* **2016**, *46*, 647–658. [CrossRef]

125. Gaskell, G.; Allum, N.; Stares, S. A Report to the EC Directorate General for Research from the Project "Life Sciences in European Society". *Eur. Biotechnol. 2002* **2003**, *58*.

126. O'Connor, E.; Cowan, C.; Williams, G.; O'Connell, J.; Boland, M.P. Irish consumer acceptance of a hypothetical second-generation GM yogurt product. *Food Qual. Prefer.* **2006**, *17*, 400–411.

127. Bech-Larsen, T.; Grunert, K.G. Can health benefits break down Nordic consumers' rejection of genetically modified foods? A conjoint study of Danish, Norwegian, Swedish and Finnish consumers preferences for hard cheese. In Proceedings of the ANZMAC 2000 Visionary Marketing for the 21st Century Conference, Griffith, Australia, 28 November–1 December 2000; pp. 78–82.

128. Honkanen, P.; Verplanken, B. Understanding attitudes toward genetically modified food: the role of values and attitudes strength. *J. Consum. Policy* **2004**, *27*, 401–420. [CrossRef]

129. Szczurowska, T. "Poles on biotechnology and genetic engineering", TNS OBOP, Plant Breeding and Acclimatization Institute, Radzikow. 2015. Available online: www.ihor.edu.pl (accessed on 29 November 2015).

130. Janik-Janiec, B.; Twordowski, T. "The social acceptance of Biotechnology in 2003 Europe and Poland", Polska Federacja Biotechnologii (PFB). 2003. Available online: www.pfb.p.lodz.pl (accessed on 29 November 2015).

131. Grunert, K.G.; Lähteenmäki, L.; Nielsen, N.A.; Poulsen, J.B.; Ueland, O.; Ström, A. Consumer perceptions of food products involving genetic modification: results from a qualitative study in four Nordic countries. *Food Qual. Prefer.* **2001**, *12*, 527–542. [CrossRef]

132. Popek, S.; Halagarda, M. Genetically modified foods: Consumer awareness, opinions and attitudes in selected EU countries. *Int. J. Consum. Stud.* **2017**, *41*, 325–332. [CrossRef]

133. Tas, M.; Balci, M.; Yüksel, A.; Yesilçubuk, N.S. Consumer awareness, perception and attitudes towards genetically modified foods in Turkey. *Br. Food J.* **2015**, *117*, 1426–1439. [CrossRef]

134. Castiglioni, M.; Faccio, E.; Veronese, G.; Mosolo, A.P.; Bell, R.C. Disturbi alimentari e costruzione del significato. *Psicol. Della Salut.* **2011**, *3*, 5–28. [CrossRef]

Brief Report

Increased Grain Amino Acid Content in Rice with Earthworm Castings

Min Huang *, Chunrong Zhao and Yingbin Zou

Southern Regional Collaborative Innovation Center of Grain and Oil Crops (CICGO),
Hunan Agricultural University, Changsha 410128, China; crzhao5@163.com (C.Z.); ybzou123@163.com (Y.Z.)
* Correspondence: mhuang@hunau.edu.cn; Tel.: +86-731-84618758

Received: 3 February 2019; Accepted: 12 March 2019; Published: 15 March 2019

Featured Application: Enhancing earthworm activity in the field may be a feasible way to increase grain amino acid content in rice.

Abstract: Enhancing the nutritional value of rice can improve the health of rice consumers. Grain amino acid content is an important nutritional component. This study was conducted to test the hypothesis that the application of earthworm castings could increase the grain amino acid content in rice. Results showed that total amino acid content in the grain of rice was significantly elevated by applying earthworm castings (17 kg m^{-2}), with an average increase of 8% across four tested rice cultivars. Application of earthworm castings had no significant effect on total nitrogen (N) content but significantly increased the ratio of amino acid to N (total amino acid content/total N content) in rice grains. The results of the present study suggest that application of earthworm castings can increase grain amino acid content in rice by improving the efficiency of the N to amino acid conversion, and highlight that further studies are required to assess the effects of earthworm castings on the amino acid metabolism in rice grains.

Keywords: amino acid content; earthworm castings; rice

---

1. Introduction

Rice feeds more than half of the global population [1], and many consumers of rice are among the world's poorest, with diets that are mainly restricted to rice as it is filling, accessible, and affordable [2]. However, rice is deficient in many nutrients [3], and thus malnutrition is common in countries where rice is the main food [4]. Therefore, it is important to enhance the nutritional value of rice to improve the health of rice consumers, especially those living in poverty.

Amino acid content is an important nutritional component on which many rice breeding programs have focused [5]. Grain amino acid content in rice is determined not only by genetic factors but also by environmental factors and management practices [6,7]. Previous studies demonstrated that there is a positive linear relationship between amino acid and nitrogen (N) contents in grains of rice [8], and that amino acid content in the rice grain can be increased by applying N fertilizer at the heading stage [6].

China is the largest producer and consumer of rice in the world, accounting for about a third of the global rice economy [9]. In the past several decades, the large input of N fertilizer across wide areas of China has caused considerable environmental costs, including soil acidification, water eutrophication, and enhanced N deposition [10–12]. Because of these effects, methods other than an artificial increase of N input are sought to increase rice grain amino acid content.

Earthworm castings, or vermicompost, are a form of fertilizer produced from earthworms as they digest soil organic matter. This excrement is enriched in mineral N [13]. Our recent study showed that application of earthworm castings could improve N uptake and utilization in rice [14]. The objective of the present study was to determine the effects of applying earthworm castings to soil on amino acid and N contents in the rice grain. We hypothesized that application of earthworm castings may increase grain amino acid content in rice.

2. Materials and Methods

A micro-plot field experiment was done in Changsha (28°11′ N, 113°04′ E), Hunan Province, China in 2017. The soil in the experimental field was a Fluvisol (FAO taxonomy) with the following chemical properties at the upper 20 cm layer: pH 5.75, 34.2 g organic matter kg^{-1}, 81.6 mg available N kg^{-1}, 34.4 mg available P kg^{-1}, and 56.7 mg available K kg^{-1}.

Thirty-two bottomless plastic boxes (length × width × height = 40 cm × 40 cm × 30 cm) were pushed into the soil of the experimental field to a depth of 20 cm to establish micro-plots. Four rice cultivars (Huanghuazhan, Liangyoupeijiu, Longliangyou 97, and Xiangliangyou 396) were grown and each micro-plot was treated with either earthworm castings (17 kg m^{-2}, EC17) or no castings (0 kg m^{-2}, EC0). The level of earthworm castings for EC17 was chosen according to the product of duration of oilseed rape-growing season (218 days) and daily production rate of earthworm castings (78 g m^{-2} day^{-1}) in a no-tillage rice-oilseed rape rotation field in Nanxian (29°21′ N, 112°25′ E), Hunan Province in 2015. The daily production rate of earthworm castings was obtained from an investigation conducted on 10 randomly selected 1-m^2 areas in the field on the first day after oilseed rape harvest. The earthworm castings (pH 7.89, 61.4 g organic matter kg^{-1}, 128 mg available N kg^{-1}, 44.2 mg available P kg^{-1}, and 254 mg available K kg^{-1}) used in the experiment were collected from several no-tillage rice-oilseed rape rotation fields in Nanxian after oilseed rape harvest in 2016. The micro-plots were arranged in a split-plot design with four replicates, using the level of earthworm castings as the main plots and cultivars as the subplots.

Rice seeds were pre-germinated and then sown in a seedbed on 10 May. Rice seedlings were transplanted on 5 June with one seedling per hill and four hills per micro-plot. Earthworm castings were applied and incorporated in the upper 20 cm of the soil on the day before transplanting. Urea was split in three applications: 6.0 g N m^{-2} at 1 day before transplanting (basal), 3.6 g N m^{-2} at 7 days after transplanting, and 2.4 g N m^{-2} at panicle initiation. Superphosphate was applied as basal at a rate of 4.8 g P_2O_5 m^{-2}. Potassium chloride (8.4 g K_2O m^{-2}) was split equally in two applications: one at basal and the other at panicle initiation. The micro-plots were flooded with a water depth of about 5 cm until 7 days before maturity. Agrochemicals were used to control diseases, insects, and weeds.

At maturity, rice grains were sampled from each micro-plot. The sampled grains were oven-dried at 70 °C to a constant weight and then hulled, ground, and sieved (100 mesh). About 1.0 g of the sieved sample was weighed, hydrolyzed, and derivatized to determine total amino acid content with high-performance liquid chromatography method [15]. Approximately 0.5 g of the sieved sample was digested by H_2SO_4-H_2O_2 method to determine the total N content. The N analysis was carried out using a segmented flow analyzer (SAN Plus, Skalar Inc., Breda, The Netherlands). The ratio of amino acid to N (total amino acid content/total N content) was calculated.

Data analysis including analysis of variance and linear regression analysis was performed using Statistix 8.0 (Analytical Software, Tallahassee, FL, USA).

3. Results and Discussion

Total amino acid content in rice grains was significantly influenced by both the level of earthworm castings and cultivar (Table 1). There was no significant interaction effect between the level of earthworm castings and cultivar on total amino acid content in the grain of rice. Averaged across the four cultivars, total amino acid content in the grain was 8% higher under EC17 than under EC0. Huanghuazhan had the highest total amino acid content in grains, more than 10% higher than that of the other three cultivars.

Table 1. Effects of earthworm casting application on total amino acid and nitrogen (N) content and on ratio of amino acid to N in grains of four rice cultivars.

Earthworm Castings	Cultivar	Total Amino Acid (mg g^{-1})	Total N (mg g^{-1})	Ratio of Amino Acid to N
EC0	Huanghuazhan	67.8 (3.52)	15.3 (0.28)	4.44 (0.232)
	Liangyoupeijiu	60.8 (2.40)	13.8 (0.05)	4.42 (0.179)
	Longliangyou 97	56.7 (3.63)	13.4 (0.22)	4.22 (0.288)
	Xiangliangyou 396	58.8 (3.40)	13.7 (0.07)	4.30 (0.250)
	Mean	61.0 (1.82)	14.1 (0.20)	4.35 (0.110)
EC17	Huanghuazhan	73.0 (3.61)	15.3 (0.07)	4.78 (0.218)
	Liangyoupeijiu	62.9 (2.49)	13.6 (0.09)	4.63 (0.198)
	Longliangyou 97	64.4 (1.67)	13.5 (0.01)	4.79 (0.125)
	Xiangliangyou 396	62.0 (2.62)	13.2 (0.11)	4.71 (0.183)
	Mean	65.6 (1.65)	13.9 (0.22)	4.73 (0.084)
Analysis of variance (*F*-value)				
Earthworm castings		4.65 *	2.20 ns	6.29 *
Cultivar		5.03 **	70.44 **	0.11 ns
Earthworm castings × Cultivar		0.34 ns	1.36 ns	0.24 ns

EC0 and EC17 represent 0 and 17 kg earthworm castings m^{-2}, respectively; Values in parentheses are SE ($n = 4$); *, significance at the 0.05 probability level; **, significance at 0.01 probability level; ns, non-significance at the 0.05 probability level.

N is a principal constitute element of amino acids. It is generally considered that amino acid content is positively linearly related to N content in the rice grain [8]. Some findings of this study are consistent with those of previous studies: Huanghuazhan, the cultivar with highest grain amino acid content, also had higher grain N content than the other three cultivars (Table 1). However, unexpectedly, no significant difference was recorded in the grain N content between the EC17 and EC0 treatments. Thus, increasing the grain N content was not the mechanism for the positive impact of applying earthworm castings on amino acid content in rice grains. Our study also showed that the EC17 treatment had a significantly higher ratio of amino acid to N than EC0 (Table 1). This result suggests that the increased grain amino acid content in rice induced by applying earthworm castings might be mostly attributable to the improved efficiency of converting N to amino acid. This could be further supported and informed by that the linear regression between grain amino acid and N contents was different under EC17 and EC0 (Figure 1). These results suggest that further investigations are required to determine the effects of earthworm castings on the metabolism of amino acid in rice grains.

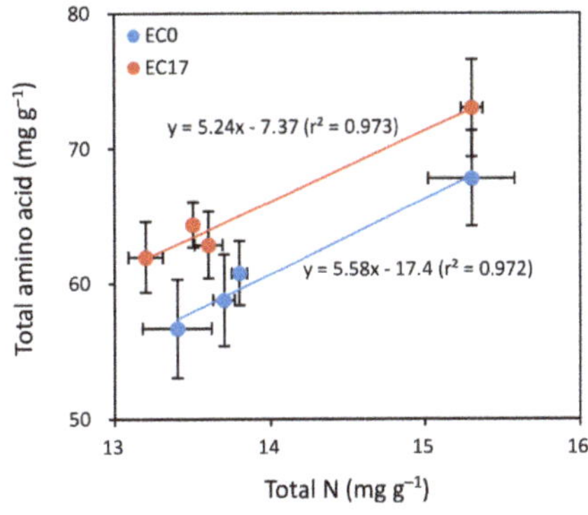

Figure 1. Linear regression between total amino acid and N contents in grains of four rice cultivars under two levels of earthworm castings. EC0 and EC17 represent 0 and 17 kg earthworm castings m^{-2}, respectively. Each data point is the mean for one cultivar under one level of earthworm castings. Error bars are SE ($n = 4$).

Application of earthworm castings can affect plant metabolism including protein synthetic activity [16,17]. In this regard, it is well documented that the castings and body secretions of earthworms contain a certain amount of plant hormones and hormone-like substances [18–20]. Earthworms can also enhance microbial biomass and activity due to their castings and mucus, and the byproducts of this microbial activity include plant hormones such as abscisic acids, auxins, cytokinins, ethylene, and gibberellins [21]. Although there is limited information available on the influences of plant hormones or plant hormone-like substances included in earthworm castings on amino acid metabolism in the rice crop, some relevant reports have been documented in other crops. For example, Muscolo et al. [22] found that earthworm-worked humic substances had an auxin-like effect on nitrogen metabolism in wild carrot, and Singh et al. [23] observed that gibberellic acid-like activity of vermicompost leachate resulted in an increase in amino acids in the common bean. Further study is required to determine if similar effects of plant hormones and hormone-like substances in earthworm castings are observed on amino acid metabolism in rice grains.

Application of earthworm castings has the potential for increasing amino acid content in rice grains, although the practicality of any approach must be taken into account. High labor costs are the key factor limiting the adoption of organic manures. In this study, the tested earthworm castings were not specially produced but were collected from no-tillage rice-oilseed rape rotation fields, where earthworms are known to be abundant (pers. obs.). Reduced or absent soil tillage in these fields can provide earthworms with an undisturbed biotope and hence may be favorable to increasing their populations, while oilseed rape plants can provide abundant food for earthworms by producing large volumes of residues. In addition, our previous study documented that adoption of no-tillage rice-oilseed rape cropping systems can maintain crop yields while saving labor [24]. Therefore, development of similar no-tillage systems may be an indirect but feasible way to increase grain amino acid content in rice by increasing earthworm activity.

Author Contributions: Conceptualization, M.H. and Y.Z.; Investigation, C.Z.; Writing, M.H.; Funding Acquisition, M.H. and Y.Z.

Funding: This research was funded by [the National Key R&D Program of China] grant number [2016YFD0300509] and [the Earmarked Fund for China Agriculture Research System] grant number [CARS-01].

Conflicts of Interest: The authors declare no conflict of interest.

References

1. Muthayya, S.; Sugimoto, J.D.; Montgomery, S.; Maberly, G.F. An overview of global rice production, supply, trade, and consumption. *Ann. N. Y. Acad. Sci.* **2014**, *1324*, 7–14. [CrossRef] [PubMed]

2. Bhullar, N.K.; Grussem, W. Nutritional enhancement of rice for human health: The contribution of biotechnology. *Biotechnol. Adv.* **2013**, *31*, 50–57. [CrossRef]

3. Birla, D.S.; Malik, K.; Sainger, M.; Chaudhary, D.; Jaiwal, R.; Jaiwal, P.K. Progress and challenges in improving nutritional quality of rice (*Oryza sativa* L.). *Crit. Rev. Food Sci. Nutr.* **2017**, *57*, 2455–2481. [CrossRef] [PubMed]

4. Muthayya, S.; Rah, J.H.; Sugimoto, J.D.; Roos, F.F.; Kraemer, K.; Black, R.E. The global hidden hunger indices and maps: An advocacy tool for action. *PLoS ONE* **2013**, *8*, e67860. [CrossRef]

5. Wang, L.; Zhong, M.; Li, X.; Yuan, D.; Xu, Y.; Liu, H.; He, Y.; Luo, L.; Zhang, Q. The QTL controlling amino acid content in grains of rice (*Oryza sativa*) are co-localized with regions involved in the amino acid metabolism pathway. *Mol. Breed.* **2008**, *21*, 127–137. [CrossRef]

6. Dou, Z.; Tang, S.; Li, G.; Liu, Z.; Ding, C.; Chen, L.; Wang, S.; Ding, Y. Application of nitrogen fertilizer at heading stage improves rice quality under elevated temperature during grain-filling stage. *Crop Sci.* **2017**, *57*, 2183–2192. [CrossRef]

7. Liu, Q.; Wu, X.; Ma, J.; Li, T.; Zhou, X.; Guo, T. Effects of high air temperature on rice grain quality and yield under field condition. *Agron. J.* **2013**, *105*, 446–454. [CrossRef]

8. Mossé, J.; Huet, J.C.; Baudet, J. The amino acid composition of rice grain as a function of nitrogen content as compared with other cereals: A reappraisal of rice chemical scores. *J. Cereal Sci.* **1988**, *8*, 165–175. [CrossRef]

9. World Rice Statistics Database. Available online: http://ricestat.irri.org:8080/wrsv3 (accessed on 31 January 2019).

10. Guo, J.H.; Liu, X.J.; Zhang, Y.; Shen, J.L.; Han, W.X.; Zhang, W.F.; Christie, P.; Goulding, K.W.T.; Vitousek, P.M.; Zhang, F.S. Significant acidification in major Chinese croplands. *Science* **2010**, *327*, 1008–1010. [CrossRef] [PubMed]

11. Le, C.; Zha, Y.; Li, Y.; Sun, D.; Lu, H.; Yin, B. Eutrophication of lake waters in China: Cost, causes, and control. *Environ. Manag.* **2010**, *45*, 662–668. [CrossRef] [PubMed]

12. Liu, X.; Zhang, Y.; Han, W.; Tang, A.; Shen, J.; Cui, Z.; Vitousek, P.; Erisman, J.W.; Goulding, K.; Christie, P.; et al. Enhanced nitrogen deposition over China. *Nature* **2013**, *494*, 459–462. [CrossRef]

13. Parkin, T.B.; Berry, E.C. Nitrogen transformations associated with earthworm casts. *Soil Biol. Biochem.* **1994**, *26*, 1233–1238. [CrossRef]

14. Huang, M.; Zhou, X.; Xie, X.; Zhao, C.; Chen, J.; Cao, F.; Zou, Y. Rice yield and the fate of fertilizer nitrogen as affected by addition of earthworm casts collected from oilseed rape fields: A pot experiment. *PLoS ONE* **2016**, *11*, e0167152. [CrossRef]

15. Cohen, S.A.; De Antonis, K.M. Applications of amino acid derivation with 6-aminoquinolyl-N-hydroxysuccinimidyl carbamate: Analysis of feed grains, intravenous solutions and glycoproteins. *J. Chromatogr. A* **1994**, *661*, 25–34. [CrossRef]

16. Galli, E.; Tomati, U.; Grappelli, A.; Di Lena, G. Effect of earthworm casts on protein synthesis in *Agaricus bisporus*. *Biol. Fertil. Soils* **1990**, *9*, 290–291. [CrossRef]

17. Tomati, U.; Galli, E.; Grappelli, A.; Di Lena, G. Effect of earthworm casts on protein synthesis in radish (*Raphanus sativum*) and lettuce (*Lactuga sativa*) seedlings. *Biol. Fertil. Soils* **1990**, *9*, 288–289. [CrossRef]

18. Atiyeh, R.M.; Lee, S.; Edvards, C.A.; Arancon, N.Q.; Metzger, J.D. The influence of humic acids derived from earthworm-processed organic wastes on plant growth. *Bioresour. Technol.* **2002**, *84*, 7–14. [CrossRef]

19. Krishnamoorthy, R.V.; Vajranabhaian, S.N. Biological activity of earthworm cats. An assessment of plant growth promoter or levels in the casts. *Proc. Indian Acad. Sci. (Anim. Sci.)* **1986**, *95*, 341–351. [CrossRef]

20. Suthar, S. Evidence of plant hormone like substances in vermiwash: An ecologically safe option of synthetic chemicals for sustainable farming. *Ecol. Eng.* **2010**, *36*, 1089–1092. [CrossRef]

21. Xu, C.; Mou, B. Vermicompost affects soil properties and spinach growth, physiology, and nutritional value. *HortScience* **2016**, *51*, 847–855. [CrossRef]

22. Muscolo, A.; Bovalo, F.; Gionfriddo, F.; Nardi, S. Earthworm humic matter produces auxin like effects on *Daucus carota* cell growth and nitrogen metabolism. *Soil Biol. Biochem.* **1999**, *31*, 1303–1313. [CrossRef]

23. Singh, S.; Kulkarni, M.G.; Staden, J.V. Biochemical changes associated with gibberellic acid-like activity of smoke-water, karrikinolide and vermicompost leachate during seedling development of *Phaseolus vulgaris* L. *Seed Sci. Res.* **2014**, *24*, 63–70. [CrossRef]

24. Huang, M.; Zou, Y.; Feng, Y.; Cheng, Z.; Mo, Y.; Ibrahim, M.; Xia, B.; Jiang, P. No-tillage direct seeding for super hybrid rice production in rice-oilseed rape cropping system. *Eur. J. Agron.* **2011**, *34*, 278–286. [CrossRef]

Article

Wheat Bread with Dairy Products—Technology, Nutritional, and Sensory Properties

Carla Graça, Anabela Raymundo and Isabel Sousa *

LEAF—Linking Landscape, Environment, Agriculture and Food, Instituto Superior de Agronomia, Universidade de Lisboa, Tapada da Ajuda, 1349-017 Lisboa, Portugal; lopesgraca.carla@gmail.com (C.G.); anabraymundo@isa.ulisboa.pt (A.R.)
* Correspondence: isabelsousa@isa.ulisboa.pt; Tel.:+351-21-365-3246

Received: 2 September 2019; Accepted: 27 September 2019; Published: 1 October 2019

Featured Application: Bakery industry, as nutritional and functional breads, with a considerable contribution to balance the daily diet for children and seniors in terms of proteins and minerals.

Abstract: As the relation between diet and health became a priority for the consumers, the development of healthy foods enriched with functional ingredients increased substantially. Dairy products represent an alternative for new products and can be used to enhance the functional and nutritional value of bakery products. The addition of yoghurt and curd cheese to wheat bread was studied, and the impact on the dough rheology, microstructure, bread quality, and sensory properties were evaluated. Dairy product additions from 10 to 50 g and higher levels up to 70 g of yoghurt and 83 g of curd cheese were tested. Replacements were performed on wheat flour basis and water absorption. It was observed that the yoghurt additions had a positive impact on the rheology characteristics of the dough. For curd cheese additions, the best dough evaluated on extension was the 30 g of wheat flour formulation. In both cases, the microstructure analysis supported the results obtained for doughs and breads. These breads showed a significant improvement on nutrition profile, which is important to balance the daily diet in terms of major and trace minerals and is important for health-enhancing and maintenance. Good sensorial acceptability for breads with 50 g of yoghurt and 30 g of curd cheese was obtained.

Keywords: dairy products; gluten network; rheology; nutrition profile; wheat bread

---

1. Introduction

Bakery products are staple foods, widely consumed in large quantities worldwide, with an important role in human nutrition [1]. Due to the increased awareness of health issues, the bakery industry is moving to provide functional and healthy foods, mainly via fortification with satiating and active ingredients, such as proteins, fibres, minerals, vitamins, and bioactive peptides [2] in response to an increasingly demanding consumer.

The incorporation of ingredients that exhibit functional properties, in addition to traditional nutrients, is an interesting alternative to the development of innovative bakery foods. Brazilian bread cheese is known worldwide, but the incorporation of dairy products (such as yogurt and curd cheese) in bread formulations is not common in the bakery markets.

Nutritional benefits of dairy products (DP) include increasing the amount of minerals with good assimilability (mainly Ca and P), vitamins (A and B12), protein, and essential amino acids (lysine, methionine, and tryptophan). Technology benefits may also be considered and can include improvement of dough handling properties and bread quality (flavour, crumb structure, and texture). These benefits result from the effect of protein and milk fat on the bread structure [3].

Good nutrition, especially adequate, easily digestible protein and mineral intake, is a determinant factor for the human health. Protein and minerals are considered key nutritional components in a well-balanced diet with an important contribution to maintain muscle mass and bone structure [4]. Some studies have reported the technological properties of dairy products as potential ingredients into a wide variety of products [5], such as infant formulas [6], that could aid in achieving dietary requirements for children and seniors.

Yoghurt (Yg) is a fermented milk product that consists of a casein network formed at the isoelectric point [7], being the acid obtained by the activity of the specific lactic acid bacteria (LAB), *Streptococcus thermophilus* and *Lactobacillus delbrueckii ssp. bulgaricus* cultures. Yg is considered the most popular dairy product worldwide for its nutritional and health benefits, since it is a rich source of protein (casein), vitamins (B2, B6, and B12), and minerals (such as Ca, P, and K), and contributes to good microbiota. This product is a potential ingredient for bakeries, representing an interesting alternative for new bakery products [8], and can be incorporated into bread formulations as a fresh product.

Curd cheese (Cc) is a co-product obtained by the thermal denaturation and subsequent precipitation of the soluble whey proteins. These products are considered to have a high protein nutritional value, representing of 20–30% of the proteins present in bovine milk. This complex mixture of globular protein molecules (α-lactalbumin and β-lactoglobulin) presents a higher protein efficiency ratio than wheat proteins [9] and is considered to be an important source of essential amino acids (leucine, isoleucine, and valine). Previous work has shown significant increments in protein, ash, and mineral contents (Ca, K, Mg, and P) in bread with whey protein concentrate [10]. However, some authors reported that whey protein exerted some negative effects on bread quality by depressing the loaf volume and increasing the crumb firmness [11].

The aim of the present study was to evaluate the influence of yoghurt and curd cheese addition as potential ingredients for the design of new bread formulations and assess their effect on wheat dough (WD) rheology and bread quality, as well as the aging kinetics and nutritional profile.

Different contents of yoghurt and curd cheese were added to wheat flour dough, and the impact on dough rheology was assessed by small amplitude oscillatory measurements (SAOS) and extension properties. Dough microstructure was also evaluated. The breads produced were also characterized in terms of microstructure, texture profile, and shelf life. Sensory analysis was performed to evaluate the acceptability of the developed breads. The maximum addition of each dairy product was discussed in terms of the potential applications for the market.

2. Materials and Methods

2.1. Raw Material

Bread was prepared using commercial wheat flour—WF (EspigaT65) purchased from the Granel cereal milling industry (Alverca, Portugal) with (per 100 g) 13.5 g of moisture, 11.3 g of protein, and 0.6 g of ash (supplier Near Infra-Red results (Hägersten, Sweden)

The fresh yogurt (Yg) used was a commercial product from LongaVida, Portugal, with the label stating the nutritional composition per 100 g: 3.7 g of protein, 0.8 g of ash, 3.7 g of lipids, 5.5 g of carbohydrates, 0.06 g of salt, and 0.12 g of Ca. The dry extract of Yg was determined from the standard Portuguese method: NP.703-1982, corresponding to 11.5% of dry matter (88.5% of moisture).

Fresh curd cheese (Cc) from Lacticínios de Paiva (Lamego, Paiva, Portugal), was used with the following nutritional composition per 100 g: 10.7 g of protein, 3.5 g of ash, 0.2 g of fibber, 9.0 g of lipids, 3.0 g of carbohydrates, 0.2 g of salt, and 0.3 g of Ca. The dry extract of Cc was determined from the standard Portuguese method: NP.3544-1987, corresponding to 25.8% of dry matter (74.2% of moisture).

Commercial white crystalline saccharose (Sidul, Santa Iria de Azóia, Portugal), sea salt (Vatel, Alverca, Portugal), baker's dry yeast (Fermipan, Lallemand Iberia, SA, Setubal, Portugal), and SSL-E481-sodium stearoyl-2 lactylate (Puratos, Portugal) were also used.

2.2. Bread Dough Preparation

Bread dough preparation was performed according to a previous study [12]. Water absorption (WA) was determined and optimized to 61.8 g/100 g of wheat flour, according to the farinograph tests (AACC 54–21.02). Control dough and doughs enriched individually with Yg and Cc were prepared considering additions of 10, 20, 30, and 50 g, and the highest levels were 70 g for Yg and 83 g for Cc. The maximum incorporation of the dairy products is determined by the level of water of the wheat dough, i.e., the water absorption with a value of 61.8%. Replacements of dairy products were based on a 100 g of wheat flour basis (baker's formula), and the dry extract of each dairy product was considered to calculate the moisture coming from each dairy product addition, i.e., for 50 g of yoghurt addition the value for moisture is 44.2 g and dry matter 5.7 g, so we added 94.3 g of wheat flour and 17.6 g of water. The following other ingredients added were kept constant: salt: 1.0%; sugar: 0.6%; dry yeast: 2.4% and sodium stearoyl-2 lactylate: 0.3%. The different bread formulations tested are presented in Table 1.

Table 1. Bread formulations optimized to the control bread (CB) and different levels of yoghurt (Yg) and curd cheese (Cc) addition, considering the moisture content and dry extract of each dairy product (DP) *.

Ingredients	CB	Yg_{10g}	Cc_{10g}	Yg_{20g}	Cc_{20g}	Yg_{30g}	Cc_{30g}	Yg_{50g}	Cc_{50g}	Yg_{70g}	Cc_{83g}
Wheat flour	100.0	98.9	97.4	97.7	94.8	96.6	92.2	94.3	87.1	92.0	78.5
Deionized water	61.8	53.0	54.4	44.1	47.0	35.3	39.6	17.6	24.7	0.0	0.0
Yoghurt *	0.0	10.0	-	20.0	-	30.0	-	50.0	-	70.0	-
Curd cheese *	0.0	-	10.0	-	20.0	-	30.0	-	50.0	-	83.3

* 100 g of Yg: moisture—88.5 g; dry extract—11.5 g; * 100 g of Cc: moisture: 74.2 g; dry extract: 25.8 g.

2.3. Physical Characterisation of the Wheat Dough

2.3.1. Extension Properties

Wheat dough extension properties (WDEP) were accessed by uniaxial extension tests using an SMS/Kieffer Dough and Gluten Extensibility Rig probe coupled to the texture analyser (Texturometer TA-XTplus, Stable MicroSystems, Surrey, UK) with a load cell of 5 kg, according to the method described earlier [13] with some modifications. The pieces of dough were left to rest for 60 min at 30 °C. The measurement conditions were measure force in tension, test speed 1.0 mm·s^{-1}, distance 70 mm, and triggers force 5 g. The force required to stretch the dough sample and the displacement of the hook were recorded as a function of time. The parameters of major importance were (i) R—resistance to deformation (N), (ii) E—extensibility (mm), and (iii) deformation energy (N·mm^{-1}). The ratio number R/E (N·mm^{-1}) was also calculated and reflects the balance between the levels of elastic and viscous components of the wheat dough [14].

2.3.2. Rheology Characterisation

Fundamental viscoelastic behaviour of wheat dough (WD) after fermentation was studied by small amplitude oscillatory shear (SAOS) measurements using a controlled stress rheometer (Haake Mars III-Thermo Scientific, Karlsruhe, Germany) with an Universal Temperature Control-Peltier system to control temperature. The procedure was in agreement with the conditions previously optimized [12]. Stress sweeps were performed at 0.1 and 1.0 Hz to determine the linear viscoelastic region and select the stress to use. Frequency sweep tests, performed at 5 °C to inhibit the yeast fermentative activity, were applied to evaluate the effect of the Yg and Cc additions on dough structure. The evolution of the dough after fermentation time, expressed in terms of storage (G′) and loss (G″) moduli, were obtained, ranging the frequency from 0.001 Hz to 100.0 Hz at a constant shear stress within the previously determined linear viscoelastic region of each sample. All determinations were repeated at least three times to ensure reproducibility of the results.

2.4. Technological Characteristics of the Bread

2.4.1. Bread Texture

Bread texture was evaluated using a texturometer TA-XTplus (Stable MicroSystems, Surrey, UK) in penetration mode. Each bread sample was cut into slices with a height of 20 mm and 120 × 100 mm rectangular shape, and slices rested for 15 min before testing. An acrylic cylindrical probe with 10 mm diameter pierced 5 mm of the sample at 1 mm/s of crosshead speed, with a load cell of 5 kg, according to the method earlier described in reference [12]. Comparison of the bread texture with different Yg and Cc contents was performed in terms of firmness.

Bread staling was evaluated measuring firmness during a storage time of 96 h, and the aging kinetics of the bread was described as a function of the dairy product incorporation according to a linear Equation (1)

$$\text{Firmness} = A * \text{time} + B, \tag{1}$$

where A can be considered the aging kinetics and B the initial firmness.

2.4.2. Quality Parameters of Bread

Bread moisture was determined according to the standard method AACC 44–15.02. Water activity (a_W) variations were determined at room temperature (Hygrolab, Rotronic).

Bake loss (BL) of breads, which is defined as the amount of water and organic material (sugars fermented and CO_2 released) lost during baking [15], was calculated, according to the Equation (2).

$$\textit{Bake loss (\%)} = [(W_{bb} - W_{ab})/W_{bb}] \times 100 \tag{2}$$

where
W_{ab}: weight of the loaf after baking;
W_{bb}: weight of the loaf before baking.

Bread volume was determined by rapeseed displacement according to the standard method AACC 10-05.01 after 2h of cooling down [8]. Specific bread volume was calculated using Equation (3).

$$\text{Specific volume (cm}^3\text{/g)} = \text{bread volume (cm}^3\text{)/bread weight (g)} \tag{3}$$

Knowing the weight and the volume of the bread, the specific weight (g/cm^3) can also be estimated using the Equation (4), i.e., the reciprocal of specific bread volume.

$$\text{Specific weight (g/cm}^3\text{)} = \text{bread weight (g)/bread volume (cm}^3\text{)} \tag{4}$$

All the experiments were done at least in triplicate.

2.5. Image Analysis of the Bread Slices

The gas cell number of the crumb breads was evaluated 120 min after production using image analysis technology. Images of three slices of breads (control, yoghurt, and curd cheese breads) were scanned full-scale using an Image Scanner (Xerox Corporation, Webster, NY, USA).

A threshold method was used for differentiating gas cells, and the segmentation was performed manually by binarization of grey-scale images into black-and-white images using the Otsu and default ImageJ algorithms, carried out in the ImageJ-based Fiji 1.46 software package [16].

2.6. Nutritional Composition of the Bread

Total nitrogen content was analysed by the MicroKjeldahl method (ISO 20483:2006). The quantification was performed by molecular absorption spectrophotometric equipment (Skalar-San^{plus} System, Auto-Analyser, Giesen, Germany) [17,18].

Fat content was determined according to NP 4168. Ash content was determined by incineration at 550 °C in a muffle furnace (AACC Method 08-01.01).

Total mineral contents were determined by inductively coupled plasma atomic emission spectrometer (ICP-AES: Thermo System, ICAP-7000 series). All the experiments were done in triplicate.

2.7. Dough and Bread Microstructure

A scanning electron microscope (TM3030 [PLUS]—TabletUp Microspcope - Hitachi, Fukuoka Japan) was used to observe the microstructure of the control dough, doughs, and breads obtained with dairy product additions. Samples were placed on the specimen holder, dried automatically by the equipment, and the freezing model was applied (−14 °C). The observations were analysed at 400 × of magnification, and 150 × of magnification, with scale bar of 200 μm for doughs and 500 μm for bread crumb, respectively.

2.8. Sensory Evaluation

Sensory evaluation of breads was carried out in a sensory evaluation room, in individual cabinets, under white light, and at room temperature by a panel of 25 untrained panelists, ages between 20–50, who were regular consumers of bread. Samples were analyzed 24 h after baking. Before analysis, the samples were sliced into equally sized slices (2 cm thick), coded, and then randomly served. Breads were evaluated based on their appearance and sensory acceptance, scoring with the descriptors: color, aroma, taste, texture, and overall acceptability on a five-point hedonic scale, where 1—"dislike extremely"; 3—"indifferent"; 5—"like extremely". Breads were considered acceptable if their mean scores for overall acceptance were above 3 (like or like extremely).

2.9. Statistical Analysis

The experimental results were statistically analysed by determining the average value, standard deviation, and the significance level was set at 95% for each parameter evaluated. Statistical analysis (RStudio, Version 1.1.423) was performed by variance analysis, the one factor (ANOVA), and post-hoc comparisons (Tukey test).

3. Results and Discussion

3.1. Wheat Dough Extension Properties (WDEP)

The effect of yoghurt (Yg) and curd cheese (Cc) addition on wheat dough extension properties at 30 °C after 60 min of fermentation time was studied, and the results obtained are presented in Figure 1. It is clearly observed that the Yg and Cc have different impacts on dough extension properties, most probably due to the different nature and structure of the proteins added—acid precipitated caseins from Yg and soluble whey protein precipitated by thermal denaturation from Cc. The extension properties are positively affected by Yg addition and are adversely affected by Cc incorporations.

Considering Figure 1A, the dough resistance values were similar to control dough up to 50 g of Yg addition, whereas for 70 g a reduction of about 21% was observed compared to control.

For Cc additions, all the levels evaluated result in dough resistance values lower than the control, with a significant reduction of about 64% ($p < 0.05$) for the highest level of Cc added (83 g).

From Figure 1B,C, it can be seen that, at 20 g of Yg addition, a steep increase both in extensibility (B) and deformation energy (C) is observed, and remained constant for the rest of the levels evaluated, with higher values compared to the control. Relative to Cc additions, lower values of extensibility and deformation energy than the control dough were observed, except at 30 g of addition, which showed values similar to the control.

In terms of R/E ratio values, the differences observed for Yg and Cc additions are clearly reflected on Figure 1D. The resistance versus extensibility is an important ratio to evaluate the balance between dough resistance (elasticity) and extensibility (viscosity), and a good combination of both parameters is required to obtain desirable dough properties and bread quality [13].

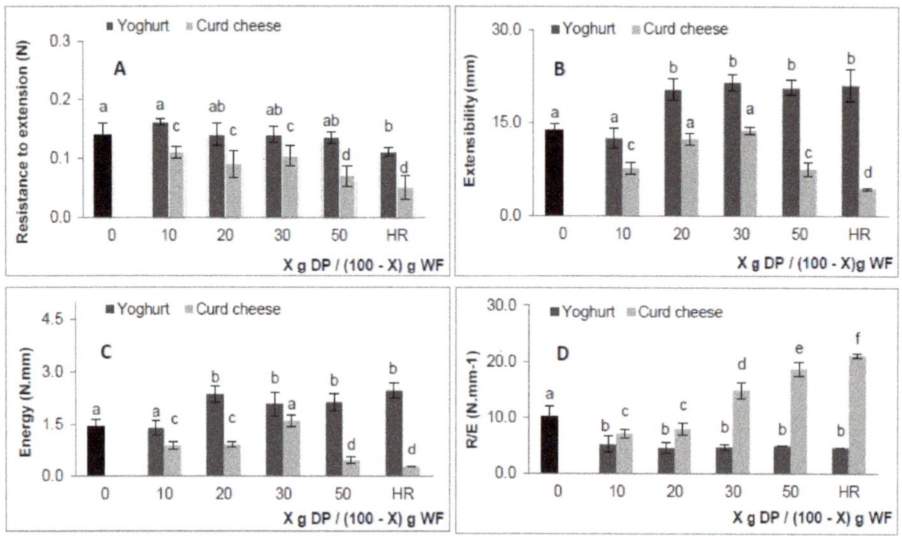

Figure 1. Effect of the Yg and Cc incorporation, after 60 min of fermentation time at 30 °C, on the extension properties of the dough, in terms of: (**A**)—resistance to extension (N), (**B**)—extensibility (mm), (**C**)—deformation energy (N.mm^{-1}) and (**D**)—ratio R/E (N.mm^{-1}), with different levels of dairy product (DP) additions: 10, 20, 30, 50, and higher replacements (HR) tested—70 g for Yg and 83 g for Cc—compared with control dough (0 g/100 g). Different letters indicate statistically significant differences at p < 0.05 (in Tukey test), compared with the control dough parameters (black bars).

Overall, the addition of Yg has a positive impact on the structure of the dough, which was seen by the increase of the extensibility and deformation energy. One can suggest a synergic interaction between the Yg proteins and wheat proteins on the gluten network development. In addition, the presence of the exopolysaccharides produced by the lactic acid bacteria present in yoghurt, can also act on the system as lubricants, together with gliadins, giving more stability and extensibility to the wheat dough. Similar results were observed by other authors [3,10] who also reported an increase in deformation energy due to the strengthening effect of hydrolyzed caseins and sodium caseinates from the added dairy product.

However, opposite results were obtained for Cc addition that affected the dough properties in extension, suggesting that denatured whey protein has an antagonistic effect with wheat proteins, strongly affecting the gluten matrix. These results can be attributed to the gluten proteins dilution effect [11,12] and protein competition for available water, impacting negatively on dough development [19]. Similar results were previously reported by other researchers [10], demonstrating that dairy by-products such as whey protein, in general, reduce the extensibility and the deformation energy.

Therefore, it can be stated that the Yg addition had no adverse impact on the technology characteristics of the dough, which will have a positive impact on bread quality. The incorporation of the Cc implied a significant reduction on extensibility properties, which would promote a significant depression on bread volume.

Dough Viscoelastic Behaviour

The changes of dough viscoelastic properties after 60 min of fermentation time were monitored by a controlled stress rheometer. The variation of storage (G´) and dissipative (G″) moduli was recorded by frequency sweeping from 0.001 to 100 Hz at a low constant stress value under linear conditions, and the G′ values were plotted at 0.1 Hz and 1.0 Hz to show the rheology changes observed by addition of the Yg and Cc to the wheat dough (Figure 2A1,B1).

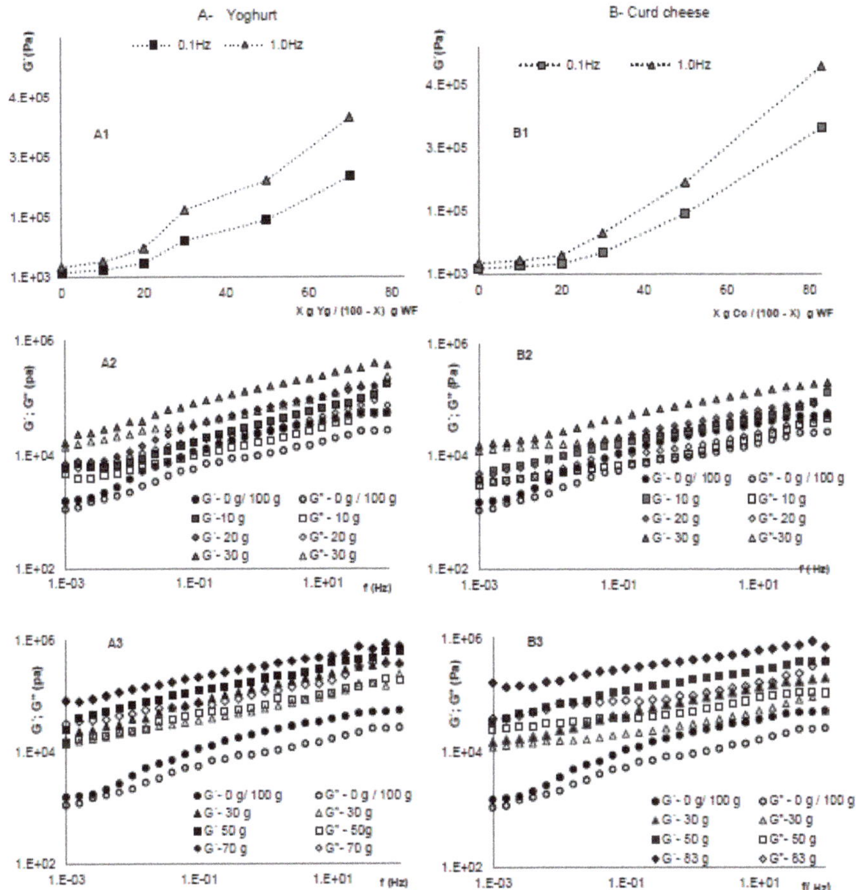

Figure 2. Variation of storage (G′) at 0.1 Hz and 1.0 Hz (**A1–B1**) and frequency sweep curves—change of the storage (G′—close symbols) and loss (dissipative (G″)—open symbols) moduli with frequency (**A2, A3; B2, B3**), for samples with differents levels of DP: A yoghurt doughs and B for curd cheese doughs.

As can be seen at Figure 2A1,B1, which are focused on G′ at 0.1 and 1.0 Hz, the behaviour of the doughs with Yg and with Cc were similar. In both, the steepest increase on slope arises from concentrations over 20 g of addition. The almost linear increase of G′ with Yg and Cc addition is characteristic of the dough structuring, indicating a reinforcement of higher density of the molecular links.

The presence of vegetable (cereals) and animal (milk) proteins adds complexity to the protein matrix with protein–protein interactions, which leads to an increasingly structured material [3,20].

Figure 2A2/A3,B2/B3 represent the mechanical spectra obtained with different Yg (A) and Cc (B) additions, where it can be observed that the G′ and G″ moduli increase with dairy product addition. In general, all the systems show a certain degree of dependence on the frequency range applied, and the addition of Yg and Cc promote the reinforcement of the dough structure, expressed in terms of viscoelastic functions (G′ and G″) increase.

The apparent disagreement between technology performances—empirical large amplitude extensional rheology (Kieffer Dough and Gluten Extensibility Rig) and small amplitude oscillatory fundamental rheology—can be explained, beyond the different nature and structure of the proteins in play, if the presence of a significant amount of exopolyssacarides in Yg is considered. In fact, the major differences in nature from Yg and Cc are the content of exopolyssacarides produced by lactic acid bacteria and protein structural characteristics of the Yg material. As was recently published [21], the impact of exopolyssacarides on the rheology of dough shows positive interactions with the gluten matrix development.

3.2. Evaluation of Bread Properties

3.2.1. Bread Texture and Aging Kinetics

The texture of the breads produced with different contents of yoghurt and curd cheese was evaluated in terms of firmness by a puncture test. The determination of firmness during a storage time of 96 h at room temperature was also performed to evaluate the effect of dairy product additions on the kinetics of bread aging (Figure 3A,B) [12]. From Figure 3A,B, it can be observed that both Yg and Cc additions had different impacts on the aging kinetics of the breads.

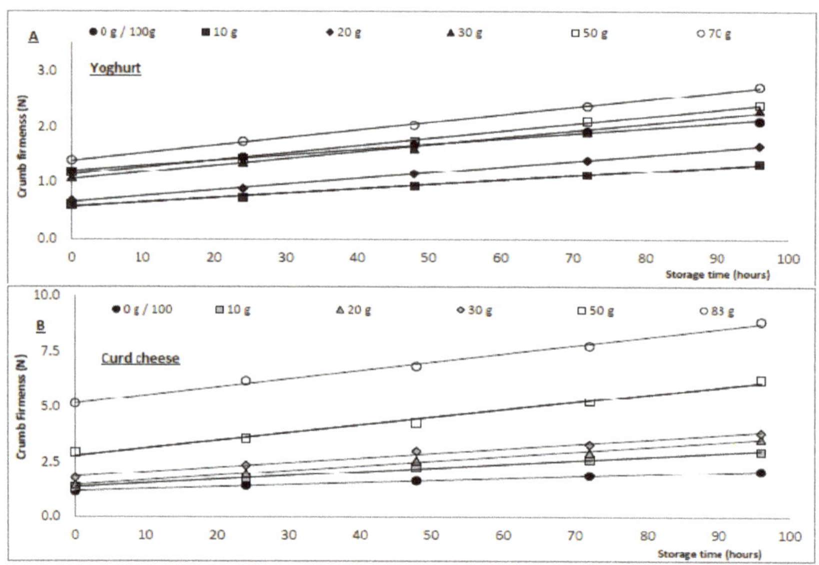

Figure 3. Variation of firmness (N) and aging kinetics for breads prepared with different contents of Yg (**A**) and Cc (**B**) addition, replacing the flour in bread formulation, at initial and 96 h of storage time, and respective linear equations: staling rate as crumb firmness with time for the different concentrations of Yg and Cc.

Bread aging kinetics was described as a function of the dairy products incorporation, and a positive linear relation ($R^2 > 0.98$) was observed, with parameters presented in Table 2, and clearly reflect the impact of the different levels of Yg and Cc addition on bread aging kinetics (A, the slope) and firmness (B, the interception).

As can be seen from Figure 3A,B and the respective linearizations on Table 2, the Yg breads are clearly less firm than Cc breads: the initial firmness of the control bread (1.21 N) is higher than low incorporations of Yg up to 50 g (1.15 N), and for higher replacements (70 g), the firmness is slightly higher (1.39 N) than control.

Table 2. Parameters of linear relationship: bread aging kinetics (A) and firmness (B) obtained for control bread (CB—0 g/100 g) and for the different levels of yoghurt (Yg) and curd cheese (Cc) tested.

DP Levels	Aging Kinetics (A)	Firmness (B)	R^2
CB	0.94×10^{-2}	1.21	0.997
Yg_{10g}	0.78×10^{-2}	0.58	0.991
Yg_{20g}	1.03×10^{-2}	0.67	0.999
Yg_{30g}	1.23×10^{-2}	1.06	0.992
Yg_{50g}	1.28×10^{-2}	1.15	0.998
Yg_{70g}	1.36×10^{-2}	1.39	0.995
Cc_{10g}	1.70×10^{-2}	1.40	0.995
Cc_{20g}	2.14×10^{-2}	1.50	0.995
Cc_{30g}	2.08×10^{-2}	1.87	0.987
Cc_{50g}	3.45×10^{-2}	2.81	0.988
Cc_{83g}	3.72×10^{-2}	5.18	0.994

The Yg breads aging kinetics (A) ranging from 0.78×10^{-2} N/h for the 10 g up to 1.36×10^{-2} N/h for the 70 g of Yg addition and are similar to control bread (0.94×10^{-2} N/h). In the case of Cc breads, a completely different effect can be observed, i.e., all breads are firmer and the rate of staling is higher than the control bread. The initial firmness up to 30 g of addition is similar to the control, but at 50 g of Cc addition, the increased is about 130%, and at 83 g, it goes up to 330%.

The rate of staling is also increasing substantially, ranging from 1.70×10^{-2} N/h for the 10 g, up to 3.72×10^{-2} N/h for the 83 g curd cheese addition, representing an increase of 120%.

Similar results were obtained in earlier studies [11], indicating that the whey protein significantly increase the bread crumb firmness. Our findings are also in agreement with those obtained by other authors [3,10] with the addition of dairy products in bread dough formulations.

This difference in bread texture with addition of Yg and Cc should reflect the complexity of the interactions of the major macromolecules at play and are in agreement with the results observed in extension properties evaluation (Figure 1). Starting from the gluten matrix, when yoghurt is added, some reinforcement of the structure was seen by the increase of extensibility and deformation energy. Values of dough resistance were similar to the control, not interfering substantially on the R/E values. The opposite is observed for curd cheese breads, where R/E values increased steeply, revealing an imbalance of the elastic and viscous components of the wheat dough, clearly reflected in bread texture and aging kinetics.

3.2.2. Quality Parameters of the Bread

The quality features of breads, i.e., moisture content, water activity (a_W), specific bread volume (SBV), and the percentage of bake loss (BL), were determined, as well as the crumb cells numbers. The results of quality parameters obtained for yoghurt and curd cheese breads are summarized in Table 3. No significant differences were observed in initial moisture content (43.5–42.6%) and water activity (0.95–0.92) for the different yoghurt breads and the control.

According to the results obtained for the Cc breads, significant differences were observed for the initial moisture content, varying between 43.5% and 38.9%, with a reduction of 11.0% of moisture, compared to the control bread and the higher Cc concentration (83 g). This effect can be explained by the water competition between wheat proteins and precipitated whey proteins during the dough development. Similar results were obtained by other authors [10] by testing the incorporation of dairy by-products on new bread formulations. Water activity had no significant variations (p > 0.05).

Table 3. Quality parameters, moisture, aw, bake loss, specific bread volume (SBV), and cell numbers of the control bread (CB—0 g/100 g) and breads produced with yoghurt (Yg) and curd cheese (Cc) addition *.

X g/(100 − X) g WF	Moisture (%)	aw	Bake Loss (%)	SBV (cm³/g)	Cell Numbers
CB	43.50 ± 0.31 [a]	0.95 ± 0.01 [a]	16.16 ± 0.58 [a]	4.44 ± 0.09 [a]	2748.5 ± 52.3 [a]
Yg₁₀g	42.81 ± 0.11 [a]	0.93 ± 0.05 [a]	13.13 ± 1.86 [a]	4.40 ± 0.25 [a]	2875.0 ± 30.1 [ab]
Yg₂₀g	43.60 ± 1.34 [a]	0.92 ± 0.02 [a]	14.70 ± 1.06 [a]	4.10 ± 0.23 [ab]	2982.0 ± 65.4 [ab]
Yg₃₀g	43.65 ± 0.34 [a]	0.93 ± 0.03 [a]	13.15 ± 2.48 [a]	3.84 ± 0.26 [ab]	3015.3 ± 38.7 [b]
Yg₅₀g	43.30 ± 0.33 [a]	0.93 ± 0.05 [a]	12.07 ± 1.29 [b]	3.61 ± 0.36 [ab]	2799.7 ± 67.0 [a]
Yg₇₀g	42.60 ± 0.27 [a]	0.92 ± 0.04 [a]	11.55 ± 2.26 [b]	3.30 ± 0.23 [b]	2568.0 ± 85.0 [a]
Cc₁₀g	44.10 ± 0.24 [a]	0.95 ± 0.03 [a]	11.40 ± 0.26 [b]	3.90 ± 0.22 [b]	1131.0 ± 77.0 [b]
Cc₂₀g	43.10 ± 0.08 [ab]	0.94 ± 0.01 [a]	10.52 ± 1.15 [b]	3.47 ± 0.18 [bc]	1035.0 ± 88.2 [b]
Cc₃₀g	42.40 ± 0.50 [bc]	0.93 ± 0.05 [a]	10.70 ± 1.41 [b]	3.21 ± 0.04 [c]	1235.0 ± 83.1 [c]
Cc₅₀g	41.93 ± 0.20 [c]	0.94 ± 0.03 [a]	9.00 ± 0.60 [b]	2.84 ± 0.16 [d]	344.0 ± 15.5 [d]
Cc₇₀g	38.85 ± 1.48 [d]	0.95 ± 0.02 [a]	9.60 ± 1.27 [b]	1.28 ± 0.09 [e]	276.0 ± 25.0 [e]

* Different letters (a, b, c, d) within the same column, for each dairy product, indicate statistically significant differences at $p < 0.05$ (Tukey test), compared with the bread control parameters.

No significant differences ($p > 0.05$) in specific bread volume (SBV) values between control and Yg breads, up to 50 g, were observed (Figure 4A), as well as in the crumb cell numbers (Table 3). The highest replacement tested, 70 g Yg, presented slightly lower SBV (3.30 cm³/g), compared to the control bread (4.44 cm³/g), representing a decrease of 26% ($p < 0.05$). In terms of bake loss values, no significant differences, up to 50 g of Yg addition, were observed. However, for 70 g of Yg addition the bake loss was lower (11.6%), compared to control bread (16.0%), and this is in line with the slightly higher density observed, leading to a less weight loss during baking.

Figure 4. Breads obtained with different levels of yoghurt (**A**) and curd cheese (**B**) incorporation: 10 g, 20 g, 30 g, 50 g and 70 g Yg, and 83 g Cc/100 g wheat flour, compared to control bread (without dairy product, 0 g/100 g wheat flour).

These good bread volumes of the yoghurt bread are probably due to the synergic effect of the different proteins on the system. In addition, the contribution of exolpolyssacarides, by building a structured polysaccharide network that interacts with the gluten matrix [21], can give more stability and extensibility to the wheat dough (Figure 1B), contributing to the gas retention and leading to a good appearance and desirable bread volume, as observed in Figure 4A.

The addition of Cc to the dough reduces significantly ($p < 0.05$) the specific bread volume (4.44–1.28 cm^3/g) during the baking process compared to control bread at the highest Cc concentration (83 g). As expected, the crumb gas cells for these breads decreased significantly ($p < 0.05$), varying from 2748.5 (control bread) to 276.0 (83 g of Cc), a reduction of 90% that is a consequence of the depression of 70% in bread volume, clearly observed in Figure 4B. These results, caused by the addition of precipitated whey protein, could be attributed to the antagonistic interaction between these proteins and the gluten complex, reducing the flexibility and the extensibility of the network (Figure 1B), which increases the density of the bread reducing the volume [1].

Our results are in agreement with those obtained by other authors [3] who reported the effect of the whey protein on loaf volume decrease, and hence on the reduction of the gas cell numbers, but sodium caseinate and hydrolyzed casein additions had no significant impact on these values.

3.2.3. Microstructure of Doughs and Breads

Environment scanning electron microscopy was used to evaluate the microstructure of the doughs, obtained with different levels of yoghurt and curd cheese (30 g and 50 g), after 60 min of fermentation at 30 °C (Figure 5A1–A5). The microstructure of the breads crumb was also observed (Figure 5B1–B5). Figure 5A1–A5 show that dough is made up of the gluten network with small and large starch granules characteristic of the wheat dough.

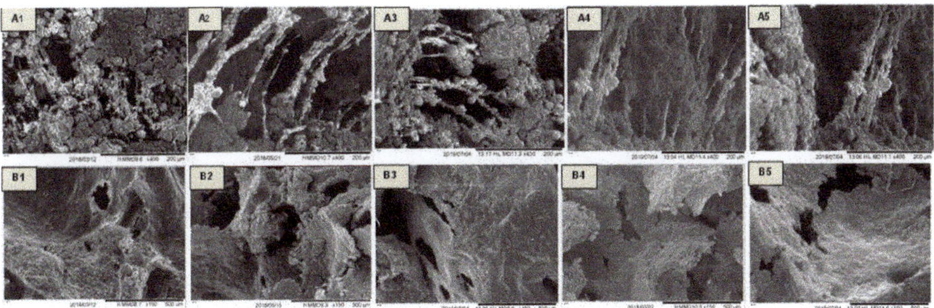

Figure 5. Scanning electron micrographs (400 ×, scale bar = 200 µm for dough; 150 ×, scale bar = 500 µm for bread crumb): A—fermented dough (60 min/37 °C); **A1**—Control dough; **A2**—30 g Yg, A3—50 g Yg; A4—30 g Cc, A5—50 g Cc; B—breads: B1—control bread, B2—30 g Yg, B3—50 g Yg; B4—30 g Cc, B5—50 g Cc, at first day of storage time.

Comparing the Yg doughs, the dough with 30 g and 50 g (Figure 5A2,A3) displayed a remarkable gluten film, with defined glutenins strands when compared to control dough, which is more evident at 30 g of Yg addition. These results are in agreement with those obtained on the extension properties, supporting the improvement of the flexibility of the net responsible for the increase on the extensibility and deformation energy values (Figure 1B,C). With Cc additions, the gluten film became less notable, as shown in Figure 5 (A4—30 g and A5—50 g of Cc addition), revealing that denatured whey protein interfered with the development of the gluten network, reducing the extensibility of the net (Figure 1B) and thus affecting the gas retention. Therefore, with the structure morphology of these doughs, the reduction of the specific volume of the breads was expected, as observed (Figure 4B).

Figure 5 (B1–B5) show the changes of crumb bread structure of the Yg bread (B2–B3) and Cc bread (B4–B5) compared to the control bread (B1). A more complex cell structure with higher number of gas cells between starch granules and denatured gluten in bread crumb with 30 g of Yg compared to the control is also noticed. This microstructural feature is associated with a lower staling rate, a low degree of firmness, and better bread volume [8]. For breads with 50 g of Yg addition, a more sheet-like structure was observed, probably due to the higher caseins interactions and most probably by the presence of exopolysaccharides that interact with the gluten network, leading to a more homogeneous structure [21]. For the Cc addition, Figure 5B4,B5, a heterogeneous and disaggregated bread crumb is observed, which is a consequence of the interference of the denatured whey protein in the gluten network.

The scanning electron microscopy images obtained for breads with Yg and Cc revealed that the interactions between dairy proteins with different nature and structures with gluten and starch matrix could partially explain and support the results obtained in dough extensibility and bread quality properties.

3.3. Nutritional Composition of the Breads

The nutritional composition, including the mineral contents, was determined for control and experimental breads obtained with 30 g and 50 g of Yg and Cc addition. A positive impact on protein and ash content was observed for both levels of dairy products tested. However, a remarkable effect was obtained for 50 g of addition in both cases, representing an increase of 7% and 30%, respectively, for Yg breads and 31% and 66% for Cc bread (Table 4). In terms of lipids, there was an increase of 28.0% for 50 g Yg addition and 163.0% of 30 g of Cc addition (the highest level to be considered in terms of bread quality). This increasing in fat is not high for the Yg breads but is considerable in the case of the Cc breads, as these are mainly saturated fatty acids from milk. This can be considered an additional restriction related with the curd cheese incorporation.

Table 4. Nutritional composition * and mineral content of breads produced with yoghurt and curd cheese: 30 g and 50 g of yoghurt (Yg) and curd cheese (Cc).

	Control Bread	Yoghurt Breads		Curd Cheese Breads	
(g/100 g)	0 g	30 g	50 g	30 g	50 g
Ash	1.75 ± 0.02 [a]	1.94 ± 0.14 [a]	2.28 ± 0.11 [b]	2.61 ± 0.02 [b]	2.90 ± 0.12 [c]
Lipids	1.92 ± 0.15 [a]	2.22 ± 0.21 [a]	2.46 ± 0.20 [ab]	5.06 ± 0.06 [c]	8.03 ± 0.16 [d]
Proteins	8.44 ± 0.15 [a]	8.50 ± 0.17 [a]	9.03 ± 0.57 [a]	9.85 ± 0.17 [b]	11.02 ± 0.32 [c]
Carbohydrates	44.30 ± 0.45 [a]	44.20 ± 0.45 [a]	43.4 ± 0.22 [a]	40.15 ± 0.80 [b]	32.38 ± 1.43 [c]
kcal	228.22 ± 1.39 [a]	230.75 ± 1.23 [a]	232.22 ± 1.83 [a]	244.73 ± 2.87 [c]	249.5 ± 2.13 [d]
Mineral Content (mg/100 g)					
Na (mg/g)	4.29 ± 0.07 [a]	4.06 ± 0.11 [b]	3.73 ± 0.28 [b]	4.74 ± 0.05 [c]	3.16 ± 3.39 [b]
K	215.32 ± 1.41 [a]	227.80 ± 4.85 [a]	257.25 ± 5.4 [b]	160.91 ± 3.39 [c]	117.58 ± 2.52 [c]
P	107.05 ± 1.99 [a]	115.80 ± 2.29 [a]	128.9 ± 1.05 [b]	144.06 ± 0.73 [c]	164.93 ± 5.84 [c]
S	96.8 ± 1.70 [a]	98.60 ± 2.33 [a]	102.01 ±1.16 [a]	117.11 ± 2.35 [b]	138.62 ± 3.07 [c]
Ca	79.76 ± 1.70 [a]	105.90 ± 2.68 [b]	120.5 ± 0.41 [b]	160.92 ± 2.44 [c]	195.33 ± 2.29 [d]
Mg	24.90 ± 0.77 [a]	25.72± 0.66 [a]	30.83 ± 0.05 [b]	26.54 ± 0.41 [a]	27.96 ± 0.40 [a]
Fe	2.09 ± 0.44 [a]	1.36 ± 0.23 [b]	1.50 ± 0.12 [b]	1.35 ± 0.14 [b]	1.17 ± 0.02 [b]
Zn	1.17 ± 0.03 [a]	1.16 ± 0.01 [a]	1.08 ± 0.01 [a]	1.24 ± 0.03 [a]	0.87 ± 0.01 [c]
Mn	0.48 ± 0.05 [a]	0.61 ± 0.02 [b]	0.68± 0.01 [b]	0.55 ± 0.03 [a]	0.63 ± 0.05 [b]
Cu	0.18 ± 0.02 [a]	0.64 ± 0.10 [b]	0.52 ± 0.08 [b]	0.19 ± 0.01 [a]	0.46 ± 0.05 [b]

* Different letters (a, b, c, d) within the same row indicate statistically significant differences at $p < 0.05$ (Tukey test), compared with the bread control parameters.

Milk and dairy products are valuable sources of minerals with a good assimilability [11] and exert several essential physiological functions in the human body. As they contain major minerals (Ca, K, Mg, and P) and trace elements (including Cu, Fe, Mn, Zn), incorporation of both dairy products promotes a significant improvement on mineral composition ($p < 0.05$) in general (Table 4).

In the breads with 50 g of addition, a significant increase was observed in Ca (Yg–51%, Cc–145%), K (Yg–20%), P (Yg–21%, Cc–54%), S (Yg–6%, Cc–43%), and Mg (Yg–24%, Cc–12%) compared to control bread. The fortification of the Yg and Cc breads with major and trace minerals is clearly noticeable, representing, in general, more than 15% of the recommended daily dosage for Ca (Yg–16%; Cc–24%), K (Yg–15%), and P (Yg–16%; Cc–18%).

For trace elements, interesting values were also noticed, especially for Cu (Yg–58%; Cc–46%) and Mn (Yg–32%; Cc–34%) (Reg. (CE), N° 1924/2006; Dir. N° 90/494 (CE)). These significant improvements will contribute to increase the minerals intake in not only daily diet of children's but also in adulthood and the elderly, where the aging process is associated with a gradual and progressive bone demineralization, along with lowered strength and physical endurance [4].

The results are supported by other researchers [11,22], which have demonstrated that the mineral and protein content of bread samples increased by the addition of different levels of whey protein, milk products, and dairy by-products.

3.4. Sensory Evaluation

The results of sensory analysis obtained for control bread and breads enriched with Yg (A) and Cc (B) are given in Figure 6. Appearance, flavor, crust color, aroma, texture, and overall acceptability of breads with 30 g and 50 g of both dairy products significantly differ in respect to control bread (0 g/100 g).

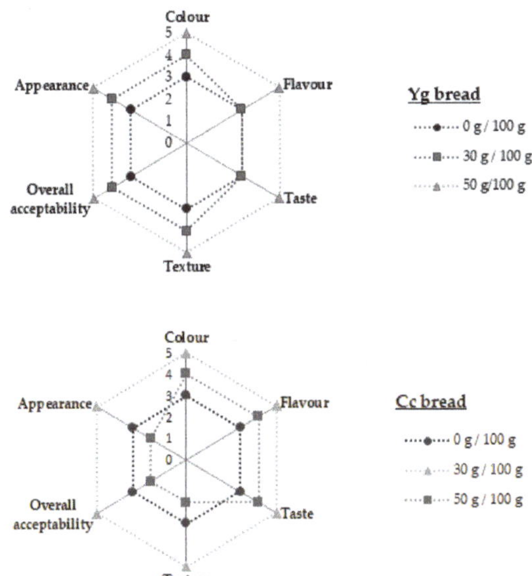

Figure 6. Sensory profile analysis of breads enriched with yoghurt (**A**) and curd cheese (**B**), with 30 g and 50 g of addition, compared with control bread (0 g/100 g).

Considering the Yg breads (Figure 6A), higher levels of yoghurt addition (50 g) caused a pleasant lactic aroma, taste, and crust color, with a significant positive influence on bread acceptability. This bread scored 5 for all attributes. In terms of texture, this Yg bread was classified as more crunchy, softer, and had a good alveoli distribution compared to the other breads (0 g and 30 g).

The best preference of Cc breads (Figure 6B) was obtained to 30 g of Cc addition, with a good acceptability in overall sensory attributes, and classified as having a better taste, crust color, and aroma. This bread also scored 5 for all attributes. For higher concentration tested (50 g), the excessive amounts of Cc proteins negatively affected the texture of this bread and the alveoli distribution, which influenced their sensorial acceptability. Results of sensory characteristics indicated that a partial replacement of wheat flour by 50 g of Yg and 30 g of Cc gave satisfactory overall consumer acceptability, and with respect to purchase intention, these breads were the best classified as "would buy for sure."

4. Conclusions

From the results of this work, it can be stated that all incorporations of the Yg resulted in good bread quality properties, and no significant adverse effect on texture and aging bread kinetics was observed. However, for Cc additions, from 30 g of Cc on, a significant increase of the crumb firmness and the aging kinetics was registered. The microstructure analysis supported the results obtained for the dough rheology and bread texture.

Considering nutritional composition, a significant increase in mineral contents was observed, indicating that the Yg and curd Cc can be used as nutritional and functional ingredients in bread formulations, which can have a significant impact on well-being. In terms of lipid content and based on higher levels to be considered as bread quality, the increase on Yg bread (50 g) was not significant, on the order of 28%, compared to control bread. However, for Cc breads (30 g), the increase in fat was much higher, around 168%, and these lipids coming from milk are mainly saturated fatty acids.

The incorporation of both dairy products studied were demonstrated to be an interesting alternative on the design of new bakery goods, with both technology properties and sensory acceptability. In terms of nutritional profile, Yg and Cc additions presented a considerable contribution to balance the daily diet, especially for children and seniors, in terms of protein, as well as in major and trace minerals, which are important to human health. However, the curd cheese addition promoted a significant increase in lipid content, which can constitute a limitation.

Author Contributions: C.G., conceived and planned the experiments; performed all samples preparation and analysis, data analysis, and interpretation of the results: and wrote the manuscript. A.R. and I.S. supervised the research work, contributed to the discussion of the data, and revised the manuscript.

Funding: This work was funded by the University of Lisbon (Doctoral Grant reference to C10781w) and by the Portuguese Foundation for Science and Technology (FCT) through the research unit UID/AGR/04129/2013 (LEAF—Linking Landscape, Environment, Agriculture and Food).

Conflicts of Interest: The authors declare no conflicts of interest.

References

1. Smith, J.P.; Daifas, D.P.; El-Khoury, W.; Koukoutsis, J.; El-Khoury, A. Shelf life and safety concerns of bakery products. *Crit. Rev. Food Sci. Nutr.* **2004**, *44*, 19–55. [CrossRef] [PubMed]

2. Fardet, A. Complex foods versus functional foods, nutraceuticals and dietary supplements: Differential health impact (Part 1). *Agro Food Ind. Hi Tech.* **2015**, *26*, 20–24.

3. Kenny, S.; Wehrle, K.; Stanton, C.; Arendt, E.K. Incorporation of dairy ingredients into wheat bread: Effects on dough rheology and bread quality. *Eur. Food Res. Technol.* **2000**, *21*, 391–396. [CrossRef]

4. Walker, S.P.; Wachs, T.D.; Grantham-McGregor, S.; Black, M.M.; Nelson, C.A.; Huffman, S.L. Inequality in early childhood: Risk and protective factors for early child development. *Lancet* **2011**, *378*, 1325–1338. [CrossRef]

5. Urista, C.M.; Fernandez, R.A.; Rodriguez, F.R.A.; Cuenca, A.A.; Tellez, A.J. Review: Production and functionality of active peptides from milk. *Food Sci. Technol. Int.* **2011**, *17*, 293–317. [CrossRef] [PubMed]

6. Li, S.; Zhang, T.; Cai, M.; Huang, B.; Ren, Y. Evaluation of amino acid conversion method for calculating whey protein content in infant formula powder. *J. Food Saf. Qual.* **2014**, *5*, 219–226.

7. Tamime, A.Y.; Robinson, R.K. *Yoghurt: Science and Technology*, 2nd ed.; CRC Press LLC: Boca Raton, FL, USA, 1999.

8. Sharafi, S.; Yousefi, S.; Faraji, A. Developing an innovative textural structure for semi-volume breads based on interaction of spray-dried yogurt powder and jujube polysaccharide. *Int. J. Biol. Macromol.* **2017**, *104*, 992–1002. [CrossRef] [PubMed]

9. Morens, C.; Bos, C.; Pueyo, M.E.; Benamouzig, R.; Gausseres, N.; Luengo, C. Increasing habitual protein intake accentuates differences in postprandial dietary nitrogen utilization between protein sources in humans. *J. Nutr.* **2003**, *133*, 2733–2740. [CrossRef] [PubMed]

10. Madenci, A.B.; Bilgiçli, N. Effect of whey protein concentrates and buttermilk powders on rheological properties of dough bread quality. *J. Food Qual.* **2014**, *37*, 117–124. [CrossRef]

11. Wronkowska, M.; Jadacka, M.; Soral-Smietana, M.; Zander, L.; Dajnowiec, F.; Banaszczyk, P.; Jelinski, T.; Szmatowicz, B. ACID whey concentrated by ultrafiltration a tool for modelling bread properties. *LWT-Food Sci. Technol.* **2015**, *61*, 172–176. [CrossRef]

12. Graça, C.; Fradinho, P.; Sousa, I.; Raymundo, A. Impact of *Chlorella vulgaris* on the rheology of wheat flour dough and bread texture. *LWT-Food Sci. Technol.* **2018**, *89*, 466–474. [CrossRef]

13. Bureova, I.; Kracmar, S.; Dvorakova, P.; Streda, T. The relationship between rheological characteristics of gluten-free dough and the quality of biologically leavened bread. *J. Cereal Sci.* **2014**, *60*, 271–275. [CrossRef]

14. Hou, G.G. *Asian Noodles: Science, Technology, and Processing*; Wiley and Sons, Inc.: Hoboken, NJ, USA, 2010.

15. Alvarez-Jubete, L.; Auty, M.; Arendt, E.K.; Gallagher, E. Baking properties and microstructure of pseudocereal fours in gluten-free bread formulations. *Eur. Food Res. Technol.* **2010**, *230*, 437–445. [CrossRef]

16. Crowley, P.; Grau, H.; Arendt, E.K. Influence of additives and mixing time on crumb grain characteristics of wheat bread. *Cereal Chem.* **2000**, *77*, 370–375. [CrossRef]

17. Krom, M. Spectrophotometric determination of ammonia; a study of modified Berthelot reaction using salicylate and dichloroisocyanurate. *Analyst* **1980**, *105*, 305–316. [CrossRef]

18. Searle, L. The Berthelot or indophenol reaction and its use in the analysis chemistry of nitrogen. *Analyst* **1984**, *109*, 549–568. [CrossRef]

19. Noisuwan, A.; Hemar, Y.; Bronlund, J.E.; Wilkinson, B.; Williams, M.A.K. Viscosity, swelling and starch leaching during the early stages of pasting of normal and waxy rice starch suspensions containing different milk protein containing different milk protein ingredients. *Starch-Stärke* **2007**, *59*, 379–387. [CrossRef]

20. Graça, C.; Raymundo, A.; Sousa, I. Rheology changes in oil-in-water emulsions stabilized by a complex system of animal and vegetable proteins induced by thermal processing. *LWT-Food Sci. Technol.* **2016**, *74*, 263–270. [CrossRef]

21. Lynch, K.M.; Coffey, A.; Arendt, E.K. Exopolysaccharides producing lactic acid bacteria: Their techno-functional role and potential application in gluten-free bread products. *Food Res. Int.* **2018**, *110*, 52–61. [CrossRef] [PubMed]

22. Bilgin, B.; Dalioglu, O.; Konyali, M. Functionality of bread made with pasteurized whey and/or buttermilk. *Ital. J. Food Sci.* **2006**, *3*, 277–286.

Article

Antihypertensive Peptide Activity in Dutch-Type Cheese Models Prepared with Different Additional Strains of *Lactobacillus* Genus Bacteria

Monika Garbowska [1,*], **Antoni Pluta** [2] **and Anna Berthold-Pluta** [2]

1   Prof. Wacław Dąbrowski Institute of Agricultural and Food Biotechnology, Inter-department Problem Group for Dairy Industries, Rakowiecka 36 Street, 02-532 Warsaw, Poland

2   Division of Milk Biotechnology, Faculty of Food Sciences, Warsaw University of Life Sciences-SGGW, Nowoursynowska 159c Street, 02-776 Warsaw, Poland; antoni_pluta@sggw.pl (A.P.); anna_berthold@sggw.pl (A.B.-P.)

*   Correspondence: garbowska-m@wp.pl; Tel.: +48-022-6063786

Received: 11 February 2019; Accepted: 4 April 2019; Published: 23 April 2019

Abstract: The objective of this study was to determine the proteolytic activity of bacterial strains from the genus *Lactobacillus* and their capability in producing peptide inhibitors of angiotensin-converting enzyme (ACE) in cheese models prepared with their addition. After 5 weeks of ripening, all cheese models studied were characterized by a high ability of angiotensin convertase inhibition which exceeded 80%. The use of the adjunct bacterial cultures from the genus *Lactobacillus* contributed to lower IC_{50} values compared with the value determined for the control cheese model. The proteolytic activity of model cheeses varied in their increase through the period of ripening, with changes in values dependent on the adjunct lactic acid bacteria (LAB) strain used for cheesemaking. Starting from the first week of ripening, the lowest proteolytic activity was demonstrated for the control cheese models, whereas the highest activity throughout the ripening period was shown for the cheese models with the addition of *Lb. rhamnosus* 489.

Keywords: ACE inhibitory activity; adjunct culture; proteolysis; cheese models

1. Introduction

Fermented dairy products are often referred to as functional products even though they are not classified in legal regulation under this name as a separate category of food products. Functional products are those which, apart from providing nutrients, contribute to an improvement in a health condition or minimize the risk of incidence of certain diseases, such as circulatory diseases, neoplasms, or osteoporosis, and also offer specified dietetic values for persons with metabolic disorders [1].

Protein fragments, which remain inactive in sequences of precursor proteins, are released upon enzymatic hydrolysis with proteolytic enzymes, and are likely to interact with respective body receptors. The proteins regulating physiological functions are called biologically and functionally active peptides or bioactive peptides, and are produced during the proteolysis of milk proteins, which provide nitrogen compounds to lactic acid bacteria (LABs) and exhibit various activities [2].

The best known currently characterized group of food-derived bioactive peptides is the group of peptides with antihypertensive properties. Most of the representatives of this group are inhibitors of the angiotensin I-converting enzyme—peptidyldipeptide hydrolase (EC.3.4.15.1), also called angiotensin-converting enzyme (ACE). ACE hydrolyzes angiotensin I to angiotensin II, which interacts with two receptors and thereby induces the contraction of blood vessels and, consequently, contributes to blood pressure increase. Hence, food products that may block the reaction of angiotensin I conversion to angiotensin II, such as ripening rennet cheeses, may be natural equivalents of hypotensive drugs. Most peptides derived from milk protein possess multi-functional properties [2–4].

The literature reports on 420 peptides with ACE-inhibitory activity (expressed as $IC_{50} < 1000$ mM) that are derived from eight species and nine proteins of milk, of which 327 are claimed to be unique peptide sequences [5]. Casein is a predominating protein, which represents 77% of all described ACE-inhibiting peptides. This results from the fact that most of the studies addressing this subject are conducted with cheeses, which are composed mainly of casein. Ripening cheeses represents an important source of bioactive peptides, obtained by multiple proteases of LABs, and of other adjunct microflora [5].

The presence of active peptides in cheeses, which are produced with natural methods, depends on the equilibrium between their synthesis and their degradation by the proteolytic system throughout the ripening process of cheeses. Peptidolytic activity is strictly related to the aging of cheese (ripening) and to the type and culture conditions of adjunct starters, as reported by different authors [6–8]. Intense proteolytic processes occurring during cheese ripening have been reported to enhance the activity of ACE inhibitors, but only to a certain level over which ACE inhibition will diminish. This may suggest that the bioactive peptides released during cheese ripening upon the activity of proteolytic enzymes of LABs are successively degraded to inactive fragments as a result of further proteolysis [9,10].

Some dairy products (Calpis and Evolus) with documented clinical effects on arterial blood pressure reduction are currently available on the market. In vivo studies are in the process of confirming the functionality of food products containing bioactive peptides in patients [2,11,12].

Some of the literature describes dairy starter cultures used for the manufacture of fermented dairy products (e.g., *L. helveticus*, *L. delbrueckii* ssp. *bulgaricus*, *L. plantarum*, *L. rhamnosus*, *L. acidophilus*, *Lc. lactis*, or *S. thermophilus*) which are capable of synthesizing bioactive peptides [13,14].

Considering the aforementioned information, a study was undertaken to determine the possibility of synthesis of bioactive peptides with hypotensive properties by selected strains of *Lactobacillus* spp. in cheese models prepared with their addition.

2. Material and Methods

The experimental material included five model cheeses produced with CHN-19 culture (*Lc. lactis* ssp. *cremoris*, *L. mesenteroides* ssp. *cremoris*, and *Lc. lactis* ssp. *lactis biovar diacetylactis*) used as the basic starter and with *Lb. casei* 2639, *Lb. acidophilus* 2499, *Lb. rhamnosus* 489, and *Lb. delbrueckii* 490 applied as additional cultures.

2.1. Determination of Proteolytic Activity of Lactobacillus Strains

Both the initial and the overall proteolytic activity of all cultures were determined using the method of Church et al. [15]. The proteolytic activity of each *Lactobacillus* strain was determined in reconstituted skim milk (RSM) after 6 and 24 h of fermentation at a temperature of 37 °C for determination of the initial and the overall proteolytic activity, respectively. All the organisms were activated from their frozen forms by one transfer into MRS broth (*Lactobacillus* Broth acc. to DE MAN, ROGOSA and SHARPE, Merck, Poland). The obtained cultures were passaged twice by a transfer 1 mL of inoculum into 100 mL of sterile supplemented (with glucose and yeast extract) reconstituted skim milk (SRSM). Then, 1 mL of the inoculum from the SRSM was transferred into 100 mL of RSM and after 24 h of incubation at 37 °C, and 1 mL of culture was transferred to 100 mL of RSM and incubated for either 6 or 24 h.

For determinations of the proteolytic activity of the analyzed strains, trichloroacetic acid (TCA) filtrates of the samples were prepared by mixing 5 mL of the sample with 1 mL of distilled water and 10 mL of 0.75 N TCA (Avantor Performance Materials, Poland), followed by centrifugation (MPW-352R centrifuge, Poland) at 4000 g and 4 °C for 30 min. The supernatants were filtered through a 0.45 μm syringe membrane filter (MILLEX HV, Milipore, Poland). The proteolytic activity of all cultures was determined by the reaction of 150 μL of the TCA filtrate with 3 mL of o-phthaldialdehyde reagent (OPA, Sigma-Aldrich, Poland). Absorbance was measured after vortexing and 2 min incubation at room temperature at 340 nm (Genesis UV-VIS Spectrophotometer, Thermo Scientific).

2.2. Preparation of Cheese Models

Cheese models were prepared in 500 mL sterile centrifuge bottles (Nalgene centrifuge ware, Thermo Scientific) [16]. 400 mL of commercial, industrially-pasteurized, and microfiltrated (74 ± 1 °C/15 s) milk (3.2% protein and 2% fat) were poured into the bottles (which were autoclaved prior to use) that were then placed in a water bath at a temperature of 35 °C. After this, the basic commercial CHN-19 starter (at a concentration 8.94 log CFU/mL) and, depending on cheese model variant, the adjunct culture (at a concentration 9.14 log CFU/mL), as well as 0.2 mL (4%) of a solution of coagulating enzymes (Fromase 2200 TL Granulate, DSM Food Specielities BV, The Netherlands) were added to the bottles. The bottles were closed and their content was mixed. The bottles were placed in a water bath with a temperature of 35 °C. After coagulation (for ca. 20–30 min), they were kept in a water bath for 20 min in order to achieve the desired consistency of the curd, and then the curd was cut using a narrow knife made of stainless steel. The curd–whey mixture was shaken for 20 min. 160 mL ± 3 mL of whey was discarded and the same volume of water with a temperature of 35 °C was added. Curd washing ended after 10 min of mixing, and afterwards the bottles were centrifuged for 10 min (320 g) at room temperature in order to remove what was likely the greatest amount of the water phase. Once the water had been removed, the bottles were centrifuged again (1400 g) at 30 °C for 1 h. The whey was then decanted and the curd was centrifuged (1400 g) for 30 min. After centrifugation, miniature cheese models were kept in bottles in a water bath (35 °C) until a pH of 5.20 was reached. Next, the cheeses were salted by pouring them in with 35 mL of saturated brine (270 g NaCl/L, Avantor Performance Materials, Poland) with a temperature of 11 °C, in the same vessel. After 5 min, the brine was removed and cheeses were transferred into sterile boxes with a grid, which facilitated whey draining, and were then placed in a cold store at 11 °C for 24 h. The weight of the cheese models were 70 ± 5 g. The prepared experimental models of cheeses (described below) were then taken out from the boxes, vacuum-packed in polyethylene foil (Cryovac packaging), and stored at 11 ± 0.5 °C for 5 weeks.

Five variants of cheese models were prepared in the study and determined as:

- C—a control cheese model consisting of milk, 2% of basic CHN-19 starter, and a coagulant;
- Lba 2499—a cheese model consisting of milk, 2% of CHN-19 starter, 1.5% of *Lb. acidophilus 2499* culture, and a coagulant;
- Lbr 489—a cheese model consisting of milk, 2% of CHN-19 starter, 1.5% of *Lb. rhamnosus 489* culture, and a coagulant;
- Lbd 490—a cheese model consisting of milk, 2% of CHN-19 starter, 1.5% of *Lb. delbrueckii 490* culture, and a coagulant;
- Lbc 2639—a cheese model consisting of milk, 2% of CHN-19 starter, 1.5% of *Lb. casei 2639* culture, and a coagulant.

Cheese model variants (C, Lba 2499, Lbr 489, Lbd 490, and Lbc 2639) were obtained in replicates performed three times.

2.3. Measurement of the ACE-Inhibitory Activity of Dutch-Type Cheese Models

The ACE-inhibitory activity of the TCA filtrates of the cheese models was assayed by the methods of Cushman and Cheung [17] and Ramchandran and Shah [18]. The assay mixture contained an hippuryl-L-histidyl-L-leucine (HHL, Sigma–Aldrich, Poland) solution (5 mM HHL in 0.1 M borate buffer containing 0.3 M NaCl, pH 8.3) used as a substrate and an ACE solution (from rabbit lung, 0.1 U mL^{-1}, Sigma Aldrich, Poland). The residue containing hippuric acid was dissolved in deionized water and the absorbance was measured (Genesis UV-VIS spectrophotometer, Thermo Scientific) at 228 nm against deionized water as a blank.

The percentage of inhibition (ACE) was calculated using the following formula:

$$\text{ACE [\%]} = \left(1 - \frac{C - D}{A - B}\right) \times 100\%$$

where:

A is the absorbance in the presence of ACE and without the sample, B is the absorbance without ACE and the sample, C is the absorbance with ACE and the sample, and D is the absorbance with the sample but without ACE.

The ACE inhibition was also expressed in terms of IC_{50}, defined as the concentration of protein in a sample ($mg\ mL^{-1}$) required to inhibit 50% of the ACE activity. The IC_{50} value was predicted by determining protein concentration in a water soluble extract (WSE) of cheese models, followed by the determination of protein concentration which ensured 50% inhibition of ACE activity. The protein concentration in the TCA filtrates was determined with the method of Lowry et al. [19].

2.4. Determination of Proteolysis of Dutch-Type Cheese Models

The reaction mixture used for determinations of the proteolytic activity in WSE of cheese models contained 3 mL of an OPA reagent (o-phthaldialdehyde, Sigma-Aldrich, Poland) and 0.15 mL of WSE. The mixture was stirred, left for 2 min at a room temperature, and subjected to absorbance measurement at a wavelength of $\lambda = 340$ nm.

2.5. Statistical Analysis

All determinations were performed 2 times. Results obtained were subjected to a statistical analysis using StatGraphics 4.1 software. One-way (ANOVA) analysis of variance was conducted. Tukey's test was applied to compare the significance of differences between mean values (as honestly significance difference (HSD)) at a significance level of $\alpha = 0.05$.

3. Results and Discussion

3.1. Proteolytic Activity of Lactobacillus Strains

Results of determinations of the initial and overall proteolytic activity of the tested LAB strains are presented in Figure 1. The initial proteolytic activity of all tested LAB strains was similar and did not differ significantly from that determined in the control sample. A balance was maintained at the early stage of fermentation between the consumption of free fatty acids and the degradation of peptides and milk proteins by LABs. In contrast, differences were observed in the overall proteolytic activity measured after 24 h. Among all analyzed LAB strains, the lowest activity was exhibited by *Lb. delbrueckii* 490, while the highest activity was exhibited by the *Lb. casei* 2639; the values of absorbance—being an indicator of the ongoing proteolysis—accounted for 0.262 and 0.512, respectively (Figure 1). The proteolytic activity of the *Lb. casei* 2639 strain was ca. twofold higher than that of *Lb. delbrueckii* 490. Proteolytic activity of the strains used as adjunct by active cultures in cheesemaking that is too high is undesirable, as it may lead to rapid proteolytic transformations during ripening and, consequently, to changes in the organoleptic traits of cheeses (such as a bitter taste and an untypical aroma).

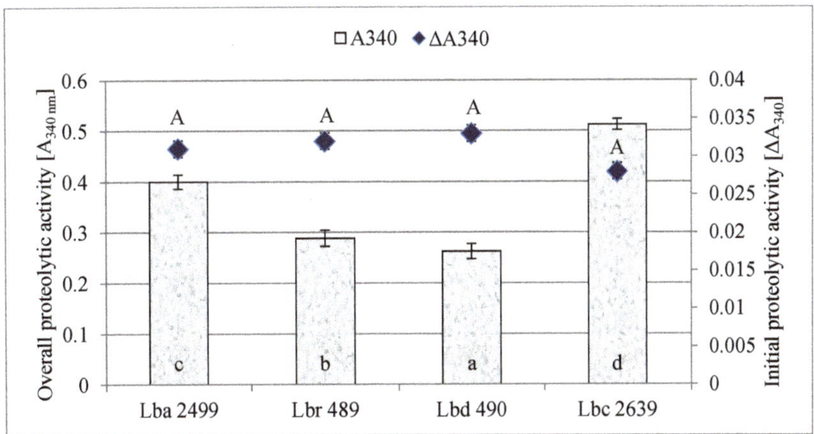

Figure 1. Initial and overall proteolytic activity of examined *Lactobacillus* strains (mean values and standard deviations). A: homogenous group initial proteolysis, n = 6; a–d: homogenous groups overall proteolysis, n = 6).

Donkor et al. [20] determined the initial and overall proteolytic activity (after 24 h incubation) of the following probiotic strains: *Lb. acidophilus* LAFTI L10 and *Lb. casei* LAFTI L26, as well as *Lb. acidophilus* 4962 and *Lb. casei* 279. Likewise, in our study we found no differences in the initial proteolytic activity between the tested strains. The lowest overall proteolytic activity was determined for the strain *Lb. casei* Lc 279 (0.290). In turn, very high activities accounting for 0.460, 0.670, and 1.90 were found for *Lb. acidophilus* LAFTI L10, *Lb. acidophilus* La 4962, and *Lb. casei* LAFTI L26, respectively. The proteolytic activity of *Lb. acidophilus* 2499 determined in our study at 0.400 is similar to that reported by Donkor et al. [20] for *Lb. acidophilus* LAFTI L10 (0.460). In turn, the activity of *Lb. casei 2639* determined in our study at 0.512 is considerably higher from the activity assayed by Donkor et al. [20] for *Lb. casei* Lc 279 (0.290) and over threefold lower from that determined for *Lb. casei* LAFTI L26 (1.90).

It was concluded that the analyzed bacterial strains from the genus *Lactobacillus* exhibited various overall proteolytic activities. Thus, large differences in proteolytic activity may result from the various activities of their proteolytic systems, and resultantly, from the various number of amine groups released in RSM during incubation, as well as from their various nutritional needs.

3.2. Proteolytic and ACE-Inhibitory Activity (%) During Ripening of Dutch-Type Cheese Models

Results of determinations of the proteolytic and ACE-inhibitory activities in the studied cheese models during their ripening are presented in Figure 2. After one week of ripening, the highest ACE-inhibitory activity was found in the cheese models with the addition of *Lb. acidophilus* 2499 (Figure 2). In other cases of cheese models, the ACE-inhibitory activity was slightly higher than in the control cheeses (69%), and ACE inhibition ranged from 74 to 79%. The control cheeses were also characterized by the lowest proteolytic activity (0.270). This slightly higher value, reaching ca. 0.305, was determined for the cheese models containing adjunct cultures *Lb. acidophilus* 2499 and *Lb. delbrueckii* 490. In turn, the highest proteolytic activity was determined for the cheese models manufactured with the addition of *Lb. rhamnosus* 489 and *Lb. casei* 2639. Similar dependencies, but with higher values, were observed in cheese models after 3 weeks of ripening, with the highest proteolytic activity being demonstrated in the cheese models with adjunct cultures *Lb. rhamnosus* 489.

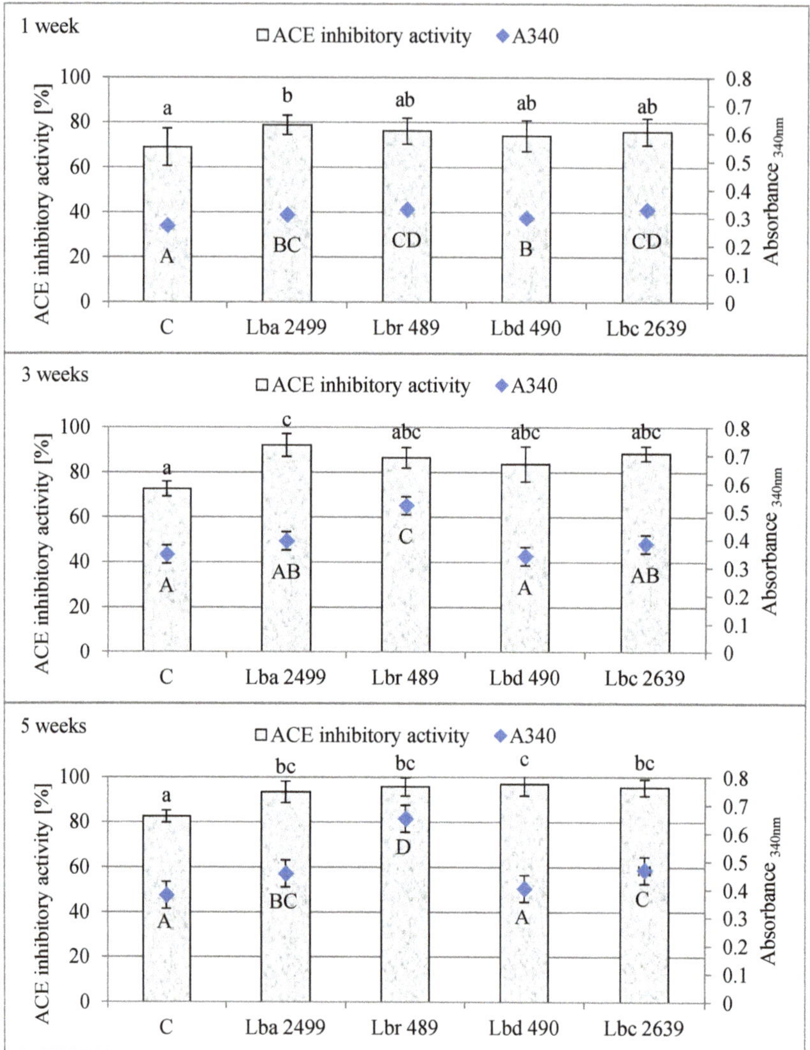

Figure 2. Angiotensin-converting enzyme (ACE) inhibitory activity (%) and proteolysis of water soluble extract (WSE) from Dutch-type cheese models during ripening (mean values and standard deviations). a–c: homogenous groups ACE, n = 6; A–D: homogenous groups proteolysis, n = 6.

One of the main factors affecting the proteolytic activity in cheeses is the water content. Water content of the analyzed cheese models (48–49%, data not shown) was higher than the typical water content of Dutch-type cheeses (42–45%). This was due to two reasons: firstly, the analyzed cheese models were not subjected to pressing and their brining was relatively short, and secondly, the higher water content of the produced models contributed to greater enhancement of proteolysis under model conditions.

After 3 and 5 week storage, a similar ACE-inhibitory activity was observed in the cheeses with adjunct cultures, despite the presence of different homogeneous groups in the Tukey tests (HSD). Cheese models with the addition of *Lb. acidophilus* 2499 were characterized by the best ACE-inhibitory capability among all cheese variants both after 1 week and 3 weeks of ripening.

After 5 weeks of storage, all analyzed cheese models were characterized by a high capability for ACE inhibition, exceeding beyond 90%. Similar ACE inhibition was noted in the cheese models with adjunct cultures of *Lb. acidophilus* 2499, *Lb. rhamnosus* 489, and *Lb. casei* 2639, however it was considerably higher than in the control cheese models. The addition of strains from the genus *Lactobacillus* to cheese models influenced the effectiveness of ACE inhibition because the tested strains contributed to a higher ACE inhibition compared to that achieved in the control cheeses at the end of ripening. The adjunct LAB strains also determined the proteolytic activity of the analyzed cheese models after 5 weeks of their ripening. The lowest proteolytic activity was determined in the cheese models with *Lb. delbrueckii* 490 and in the control models (Figure 2). A significantly higher activity compared to control cheeses was assayed in the cheese models containing adjunct cultures *Lb. acidophilus* 2499 and *Lb. casei* 2639 (ca. 0.460). In turn, the highest activity was determined in the cheese models with the addition of *Lb. rhamnosus* 489, and it was 1.5-fold higher than that determined in the control cheeses. Ong et al. [21] showed an increase in the content of inhibitors within the first 24 weeks of ripening of probiotic and control Cheddar cheeses, which remained at a similar level within the 12 subsequent weeks.

The study results indicate that the proteolytic transformations occurring during cheese model ripening are significantly influenced by the adjunct cultures which intensify casein hydrolysis by releasing peptides responsible for ACE inhibition from its chains. Considering the proteolytic activity of the analyzed LAB strains established in the cheese models by determining the number of free amine groups, being a measure of the degree of proteolysis during ripening, its values were observed to vary and increase throughout the ripening of cheese models depending on the adjunct LAB strain used in cheesemaking. There were no statistically significant differences in the cheese models with the addition of *Lb. acidophilus* 2499, *Lb. rhamnosus* 489, and *Lb. casei* 2639 cultures.

The IC_{50} values of the studied cheese models determined throughout the ripening period are presented in Table 1. Immediately after cheesemaking, the lowest IC_{50} value (0.7 mg mL^{-1}) was determined in the cheese models with the addition of *Lb. acidophilus* 2499 and *Lb. rhamnosus* 489. The addition of these two adjunct strains during cheesemaking had a significant effect on the produce of ACE inhibitors. An inconsiderably higher IC_{50} value was determined in the cheese models with the addition of *Lb. delbrueckii* 490 and *Lb. casei* 2639. In turn, the highest concentration of peptides needed to obtain 50% inhibition of ACE activity was determined in the control cheese models (0.84 mg mL^{-1}). The lowest IC_{50} values were noted in the cheese models containing lactobacilli after 5 weeks of ripening. The results above indicate that the lactobacilli produced ACE inhibitors from the beginning of the ripening process. After 5 weeks of ripening, the lowest IC_{50} value among all cheese models with the addition of LABs from the genus *Lactobacillus* was determined in those containing *Lb. delbrueckii* 490 (0.39 mg mL^{-1}), and the highest value was determined in those with *Lb. casei* 2639 (0.47 mg mL^{-1}) (Table 1). IC_{50} values were not significantly different between the control cheese model and the other three models with the addition of *Lb. acidophilus* 2499, *Lb. rhamnosus* 489, and *Lb. casei* 2639. Therefore, the low value of IC_{50} in the control cheese model may indicate a crucial role of the starter cultures in the formation of peptides with antihypertensive properties during the ripening of the examined cheese models. An increase in the ACE-inhibitory activity in cheeses containing various adjunct strains of LAB was reported by Ong and Shah [22]. A comparative analysis of the ACE-inhibitory activity of the same cheese models in our study revealed the IC_{50} parameter to be a better indicator when comparing enzyme inhibition effectiveness because it takes account of the concentration of peptides and dissolved proteins in a sample.

Table 1. IC$_{50}$ values of ACE-inhibitory activity from control cheese models and with adjunct culture (mg mL^{-1}).

	Cheese Model				
	C	Lba 2499	Lbr 489	Lbd 490	Lbc 2639
1 day	0.84 ± 0.01 d, D	0.71 ± 0.02 ab, D	0.69 ± 0.03 a, D	0.75 ± 0.06 bc, D	0.75 ± 0.02 bc, D
1 week	0.69 ± 0.05 bc, C	0.67 ± 0.03 abc, C	0.65 ± 0.02 ab, C	0.63 ± 0.06 a, C	0.67 ± 0.05 abc, C
3 weeks	0.59 ± 0.06 ab, B	0.57 ± 0.03 ab, B	0.55 ± 0.03 a, B	0.55 ± 0.06 a, B	0.57 ± 0.03 a, B
5 weeks	0.49 ± 0.05 b, A	0.45 ± 0.08 ab, A	0.45 ± 0.03 ab, A	0.39 ± 0.02 a, A	0.47 ± 0.02 ab, A

Mean values and standard deviations; a–d: means with different letters in line are significantly different ($p < 0.05$, n = 6); A–D: means with different letters in column are significantly different ($p < 0.05$, n = 6).

The results of investigations reported in the literature point to vast differences in the ACE-inhibitory activity among various cheese species and to the usability of in vitro studies in the identification of cheese samples with a high ACE-inhibitory activity. The ability to identify the stage of cheese ripening at which the concentration of bioactive peptides is the highest may help in establishing the point in the ripening process when the cheeses exhibit the greatest health-promoting properties [23,24].

The literature data indicate that the presence of ACE inhibitors is affected to a greater extent by the cheesemaking technology (including the heat treatment of milk) [10,25], starter culture and adjunct LABs used [21,22,25], and ripening conditions (period and temperature) [1,9,26–28], than by cheese species. The shelf-lives of the majority of the Dutch-type cheeses produced and consumed across the globe are not long. The recommended minimal ripening period is 5 weeks or preferably even shorter. For this reason, investigations of cheeses of this type but with a ripening period that is a few or even a dozen times longer concern a relatively low number of cheeses available in retail. In our study, the cheese models were analyzed for 5 weeks and these analyses demonstrated an increasing activity of ACE inhibitors. This confirms that the Dutch-type cheeses should be ripened for the period of 5 weeks at a minimum.

It is difficult to establish a close relationship between the ACE-inhibitory activity in vitro and the hypotensive effect in vivo. This arouses some doubts concerning the use of the ACE-inhibitory activity in vitro as the sole criterion in the identification of substances with a potentially hypotensive effect, owing to the possibility of their physiological transformations in vivo [29]. This has been confirmed in a study conducted by Bernabucci et al. [30], where the authors demonstrated that the in vitro ACE-inhibitory activity of naturally-formed bioactive peptides in Parmigiano Reggiano (PR) and Grana Padano (GP) cheeses caused no hypotensive effect in vivo. Hence, the in vitro ACE-inhibitory activity cannot be used as the sole criterion in the evaluation of potentially hypotensive substances. Therefore, it is necessary to examine the beneficial hypotensive properties of bioactive peptides in vivo considering the possibility of their enzymatic degradation or diminished absorption under these conditions.

4. Conclusions

The use of adjunct strains from the genus *Lactobacillus* increased contents of peptide inhibitors of ACE in the cheese models with their addition compared to the control cheeses. Contents of these inhibitors were observed to increase throughout 5 weeks of ripening, hence their lower amount was needed to induce 50% inhibition of ACE. Results of our study demonstrate that the adjunct strains used in cheesemaking intensified the synthesis of peptides with hypotensive properties in vitro. However, caution should be exercised regarding the obtained ACE-inhibitory activity in vitro and the speculated hypotensive effect in vivo, given that the hypotensive properties of bioactive peptides need to be confirmed under in vivo conditions.

Author Contributions: Conceptualization, M.G. and A.P.; Methodology, M.G.; Formal Analysis, M.G.; Investigation, M.G.; Writing—Original Draft Preparation, M.G.; Writing—Review & Editing, M.G., A.P. and A.B.-P.; Supervision, M.G. and A.P.

Funding: This research was funded by Wacław Dąbrowski Institute of Agricultural and Food Biotechnology, statutory subject BST [500-01-GM-01], FBW [510-01-GM-01].

Conflicts of Interest: The authors declare no conflict of interest.

References

1. Ryhänen, E.L.; Pihlanto-Leppälä, A.; Pahkala, E. A new type of ripened, low-fat cheese with bioactive properties. *Int. Dairy J.* **2001**, *11*, 441–447. [CrossRef]
2. Dziuba, B.; Dziuba, M. Milk proteins-derived bioactive peptides in dairy products: Molecular, biological and methodological aspects. *Acta Sci. Pol. Technol. Aliment.* **2014**, *13*, 5–25. [CrossRef] [PubMed]
3. Fernandez, M.; Hudson, J.A.; Korpela, R.; de los Reyes-Gavilán, C.G. Impact on human health of microorganisms present in fermented dairy products: An overview. *Biomed. Res. Int.* **2015**, *2015*, 412714. [CrossRef]
4. Sharma, S.; Singh, R.; Rana, S. Bioactive peptides: A review. *Int. J. Bioautom.* **2011**, *15*, 223–250.
5. Nielsen, S.D.; Beverly, R.L.; Qu, Y.; Dallas, D.C. Milk bioactive peptide database: A comprehensive database of milk protein-derived bioactive peptides and novel visualization. *Food Chem.* **2017**, *232*, 673–682. [CrossRef]
6. Sieber, R.; Bütikofer, U.; Egger, C.; Portmann, R.; Walther, B.; Wechsler, D. ACE-inhibitory activity and ACE-inhibiting peptides in different cheese varieties. *Dairy Sci. Technol.* **2010**, *90*, 47–73. [CrossRef]
7. Baptista, D.P.; Galli, B.D.; Cavalheiro, F.G.; Negrão, F.; Eberlin, M.N.; Gigante, M.L. *Lactobacillus helveticus* LH-B02 favours the release of bioactive peptide during Prato cheese ripening. *Int. Dairy J.* **2018**, *87*, 75–83.
8. Reale, A.; Ianniello, R.G.; Ciocia, F.; Di Renzo, T.; Boscaino, F.; Ricciardi, A.; Coppola, R.; Parente, E.; Zotta, T.; McSweeney, P.L.H. Effect of respiratory and catalase-positive *Lactobacillus casei* adjuncts on the production and quality of Cheddar-type cheese. *Int. Dairy J.* **2016**, *63*, 78–87. [CrossRef]
9. Addeo, F.; Chianese, L.; Salzano, A.; Sacchi, R.; Cappuccio, U.; Ferranti, P.; Malorni, A. Characterization of the 12% trichloroacetic acid-insoluble oligopeptides of Parmigiano-Reggiano cheese. *J. Dairy Res.* **1992**, *59*, 401–411. [CrossRef]
10. Smacchi, E.; Gobbetti, M. Bioactive peptides in dairy products: Synthesis and interaction with proteolytic enzymes. *Food Microbiol.* **2000**, *17*, 129–141. [CrossRef]
11. EFSA. Evolus and reduce arterial stiffness—Scientific substantiation of a health claim related to Lactobacillus helveticus fermented Evolus®low-fat milk products and reduction of arterial stiffness pursuant to article 14 of the regulation (EC) No 1924/2006—Scientific opinion of the panel on dietetic products, nutrition and allergies; question number EFSA-Q-2008-218. *EFSA J.* **2008**, *6*, 824. [CrossRef]
12. Korhonen, H.; Pihlanto, A. Bioactive peptides: Production and functionality. *Int. Dairy J.* **2006**, *16*, 945–960. [CrossRef]
13. Hafeez, Z.; Cakir-Kiefer, C.; Roux, E.; Perrin, C.; Miclo, L.; Dary-Mourot, A. Strategies of producing bioactive peptides from milk proteins to functionalize fermented milk products. *Rev. Artic. Food Res. Int.* **2014**, *63*, 71–80. [CrossRef]
14. Hajirostamloo, B. Bioactive component in milk and dairy product. *Int. Sci. Index Agric. Biosyst. Eng.* **2010**, *4*, 870–874.
15. Church, F.C.; Swaisgood, H.E.; Porter, D.H.; Catignani, G.L. Spectrophotometric assay using o-phthaldialdehyde for determination of proteolysis in milk and isolated milk proteins. *J. Dairy Sci.* **1983**, *66*, 1219–1227. [CrossRef]
16. Hynes, E.; Ogier, J.C.; Delacroix-Buchet, A. Protocol for the manufacture of miniature washed-curd cheeses under controlled microbiological conditions. *Int. Dairy J.* **2000**, *10*, 733–737. [CrossRef]
17. Cushman, D.W.; Cheung, H.S. Spectrophotometric assay and properties of the angiotensin-converting enzyme of rabbit lung. *Biochem. Pharmacol.* **1971**, *20*, 1637–1648. [CrossRef]
18. Ramchandran, L.; Shah, N.P. Influence of addition of Raftiline HP®on the growth, proteolytic, ACE-and α-glucosidase inhibitory activities of selected lactic acid bacteria and Bifidobacterium. *LWT Food Sci. Technol.* **2010**, *43*, 146–152. [CrossRef]
19. Lowry, O.H.; Rosebrough, N.J.; Farr, A.L.; Randall, R.J. Protein measurement with the Folin phenol reagent. *J. Biol. Chem.* **1951**, *193*, 265–275. [PubMed]
20. Donkor, O.N.; Henriksson, A.; Vasiljevic, T.; Shah, N.P. Proteolytic activity of dairy lactic acid bacteria and probiotics as determinant of growth and in vitro angiotensin-converting enzyme inhibitory activity in fermented milk. *Le Lait* **2007**, *87*, 21–38. [CrossRef]

21. Ong, L.; Henriksson, A.; Shah, N.P. Angiotensin converting enzyme-inhibitory activity in Cheddar cheeses made with the addition of probiotic *Lactobacillus casei* sp. *Le Lait* **2007**, *87*, 149–165. [CrossRef]

22. Ong, L.; Shah, N.P. Release and identification of angiotensin-converting enzyme-inhibitory peptides as influenced by ripening temperatures and probiotic adjuncts in Cheddar cheeses. *LWT Food Sci. Technol.* **2008**, *41*, 1555–1566. [CrossRef]

23. Pripp, A.H.; Sørensen, R.; Stepaniak, L.; Sørhaug, T. Relationship between proteolysis and angiotensin-I-converting enzyme inhibition in different cheeses. *LWT Food Sci. Technol.* **2006**, *39*, 677–683. [CrossRef]

24. Smacchi, E.; Gobbetti, M. Peptides from several Italian cheeses inhibitory to proteolytic enzymes of lactic acid bacteria, *Pseudomonas fluorescens* ATCC 948 and to the angiotensin I-converting enzyme. *Enzym. Microb. Technol.* **1998**, *22*, 687–694. [CrossRef]

25. Meira, S.M.M.; Daroit, D.J.; Helfer, V.E.; Corrêa, A.P.F.; Segalin, J.; Carro, S.; Brandelli, A. Bioactive peptides in water-soluble extracts of ovine cheeses from Southern Brazil and Uruguay. *Food Res. Int.* **2012**, *48*, 322–329. [CrossRef]

26. Meisel, H.; Goepfert, A.; Gunther, S. ACE-inhibitory activities in milk products. *Milchwissenschaft* **1997**, *52*, 307–311.

27. Saito, T.; Nakamura, T.; Kitazawa, H.; Kawai, Y.; Itoh, T. Isolation and structural analysis of antihypertensive peptides that exist naturally in Gouda cheese. *J. Dairy Sci.* **2000**, *83*, 1434–1440. [CrossRef]

28. Ong, L.; Shah, N.P. Influence of probiotic *Lactobacillus acidophilus* and *L. helveticus* on proteolysis, organic acid profiles, and ACE inhibitory activity of Cheddar cheeses ripened at 4, 8, and 12 °C. *J. Food Sci.* **2008**, *73*, M111–M120. [CrossRef]

29. Stuknytė, M.; Cattaneo, S.; Masotti, F.; De Noni, I. Occurrence and fate of ACE-inhibitor peptides in cheeses and in their digestates following in vitro static gastrointestinal digestion. *Food Chem.* **2015**, *168*, 27–33. [CrossRef]

30. Bernabucci, U.; Catalani, E.; Basiricò, L.; Morera, P. In vitro ACE-inhibitory activity and in vivo antihypertensive effects of water-soluble extract by Parmigiano Reggiano and Grana Padano cheeses. *Int. Dairy J.* **2014**, *37*, 16–19. [CrossRef]

applied
sciences

Article

A Novel Way for Whey: Cheese Whey Fermentation Produces an Effective and Environmentally-Safe Alternative to Chlorine

Maria Isabel S. Santos [1,2], Patrícia Fradinho [1], Sandro Martins [1], Ana Isabel G. Lima [1], Ricardo M. S. Boavida Ferreira [1], Laurentina Pedroso [2], Maria Adélia S. S. Ferreira [1] and Isabel Sousa [1,*]

[1] LEAF—Linking Landscape, Environment, Agriculture and Food, Instituto Superior de Agronomia, Universidade de Lisboa, Tapada da Ajuda, 1349-017 Lisbon, Portugal
[2] Faculty of Veterinary Medicine, Universidade Lusófona de Humanidades e Tecnologias, Campo Grande, 376, 1749-024 Lisbon, Portugal
* Correspondence: isabelsousa@isa.ulisboa.pt; Tel.: +351-21-365-3543

Received: 14 June 2019; Accepted: 10 July 2019; Published: 12 July 2019

Featured Application: Industrial waste whey, as a low-cost, efficient, and environmentally safe disinfectant, with potential applications for minimally processed foodstuff.

Abstract: Cheese whey has been described as an environmental hazard due to its high organic content. Although it has been suggested that whey can be used as food disinfectant, it continues to pose an environmental problem because it still contains a high organic load. Here, we aimed to develop a low-cost, scalable fermentation protocol to produce a disinfectant from dairy waste that has very little organic content and high levels of lactic acid. Fermentation was achieved with industrial whey from ewe, goat, and cow's milk, using a specific mesophilic-lactic acid bacteria starter mix over 120 h, which yielded the highest lactic acid production and the lowest lactose content. Antibacterial activity was observed against *Listeria monocytogenes*, *Salmonella* enterica, and *Escherichia coli* O157:H7, plus a total of thirteen other food pathogenic and spoilage strains, and antibacterial activities were determined to be highest after 120 h. We further validated this whey's application as a disinfectant in shredded lettuce and compared its efficacy to that of chlorine, evaluating microbial quality, texture, color, and sensory perception, pH, and O_2 and CO_2 determinations. Results showed that not only was microbial quality better when using our whey solution ($p < 0.05$), but also the quality indicators for whey were statistically similar to those treated with chlorine. Hence, our work validates the use of an industrial waste whey as a low-cost, efficient, and environmentally safe disinfectant, with potential applications for minimally processed foodstuffs as an alternative to chlorine.

Keywords: fermented whey; antimicrobial; minimally processed vegetables; quality markers; sensory evaluation; disinfection; chlorine alternative

---

1. Introduction

Whey is an industrial by-product resulting from cheese manufacture. Approximately 89% of the milk used for cheese manufacture is processed into whey [1]. Due to its high organic matter content, it holds a high chemical oxygen demand (COD), hence posing a considerable pollution problem [2–5], meaning that currently there are very restrictive legislations regarding whey disposal [6,7]. Hence, a novel trend for finding emerging alternative uses has risen, with one of them being the reuse of whey as a disinfecting agent [8–10]. The increasingly recognized antibacterial potential of whey is mostly due to the presence of lactic acid and antibacterial bioactive peptides [11–13], both resulting from the cheese fermentation process [14–16]. Being a food product by nature, whey has been suggested to hold particular interest as a disinfectant in freshly-cut fruit and vegetables, or minimally processed produce (MPP), which in turn are becoming major health concerns because of highly frequent pathogen outbreaks [17,18]. Indeed, several studies have reported the presence of strains such as Shiga toxin–producing *Escherichia coli*, *Salmonella* spp., and *Listeria monocytogenes* in MPP, with the last one causing death rates as high as 20% in high risk groups [17–22], followed by *Salmonella* spp., with a reported frequency of 4–8% [23,24]. Moreover, currently, chlorine is the most common disinfectant used in the fresh-cut industry [25] and it poses a serious environmental and health hazard, with strong evidences of carcinogenic problems, mostly due to the formation of toxic derivatives, such as chloramines and trihalomethanes, and restrictions to the use of chlorinated solutions are starting to arise [25–27], which consequentially has led to an increasingly higher demand to find new sanitizers [28,29]. However, as demonstrated by several published works [25,30–32], most of these alternatives are high in cost and induce odor and alterations to the organoleptic properties of the foods to which they are applied. Hence, the development of novel, cost-effective, and natural disinfectants for MPP foods is of very high economical and industrial interest. Overall, whey seems to be a promising alternative to chlorine as a disinfectant [33], and furthermore Lactic Acid Bacteria (LAB) species have been proposed as protective cultures for minimally processed food, because of their great potential for bio-control of pathogenic bacteria [34,35] and their generally recognized safe status (generally recognized as safe - GRAS, Grade One status) [36]. However, using whey or whey permeate as a disinfectant in food would result in the same environmental problem because it still contains a high load of organic matter. This could be partially overcome by fermenting the whey itself, which promotes the removal of proteins as well as lactose, and thus reduces its COD. Another direct effect of this processing is a higher production of lactic acid from lactose, which can increase its antibacterial activity, and in the case of whey proteins, it can also yield specific polypeptide sequences with antibacterial activity [37,38], which can increase its potential as a sanitizer. Nonetheless, most works use up to 24 h of fermentation [33,39], which is not enough to eliminate organic matter content. Also, it is very important to ensure that the fermented whey does not alter the organoleptic properties of the minimally processed (MP) produce, since unprocessed whey, or whey permeate, can induce visual and odor alterations to the product. However, very few works have tested whey, fermented or not, in a realistic manner in MP vegetables, particularly those more sensitive to decay, such as shredded loose-leaf lettuce. Hence, in this work, we aimed to develop a low-cost, scalable way to produce a disinfectant from dairy waste that has very little organic content and high levels of lactic acid, and induces little to no alterations in food quality.

Hence, we developed a fermentation protocol of whey from mixed origin (cow, ewe, goat) based on previous works [40,41] and evaluated its technological potential as a disinfectant in a realistic manner and in several bacterial strains. The goal was to produce the highest amount of lactic acid, whilst reducing lactose in a low-cost and efficient manner, and to determine its applications to control relevant pathogens isolated from vegetable foods, all the while maintaining food quality and safety. We also aimed to determine its efficacy against chlorine in salad disinfection. Overall, our work validated the use of an industrial waste whey as a low-cost, healthy, and efficient disinfectant that can replace chlorine, with potential applications on minimally processed foods.

2. Materials and Methods

2.1. Whey Fermentation

The whey used for fermentation was collected after cheese manufacture from a mixture of ewe, goat, and cow´s milk using an industrial starter mix (Danisco, Sassenage, France) of *Lactococcus lactis* subsp. *lactis*, *Lactococcus lactis* subsp. *cremoris*, *Lactococcus lactis* subsp. *lactis* biovar *diacetylactis*, *Streptococcus thermophilus*, and *Lactobacillus delbrueckii* subsp. *bulgaricus*, containing approximately 10^7 cfu·mL^{-1} of whey. It was kept at -18 °C until fermentation assays. Bacterial cell enumeration was performed on Man Rogosa and Sharp medium (MRS broth) (Biokar Diagnostic[TM], Beauvais, France), containing 2% (*w/v*) agar (Difco, Quilaban, Portugal). Incubation was done at 37 °C ± 1 °C for 48 h ± 2 h. The samples of whey were obtained after manufacture of cheese from a mixture of ewe, goat, and cow´s milk, containing 0.04% (*w/v*) NaCl, pH 6.65 (Lab 850, Schott AG, Mainz, Germany) and 30 g·L^{-1} lactose, quantified by HPLC (High-performance liquid chromatography) ion exchange chromatography, as in a previous work [41], was used for the fermentation assays. Whey was divided into 500 mL aliquots in Erlenmeyer flasks, and assays were conducted at 37 °C for 120 h. Over time, at regular intervals, 5 mL samples were taken for pH measurement (Lab 850, Schott AG, Mainz, Germany) and HPLC determinations, as previously described by Santos et al. [41].

2.2. Biochemical Analysis

The quantification of sugars and acids throughout the fermentation assays were done by HPLC (Waters Corporation, Milford, MA, USA), equipped with a 515 HPLC Pump (Waters Corporation, Milford, MA, USA) and incorporated with a Refractive Index Detector (RID) 486 (Waters Corporation, Milford, MA, USA). Prior to injection, samples were centrifuged at 10.000 rpm/10 min and the supernatants were filtered through a Millipore membrane with a pore size of 0.2 μm. Samples were injected in a Schodex SC-1011 column (Waters Corporation, Milford, MA, USA) and separations were achieved at 50 °C, using 5 mM sulphuric acid as mobile phase (isocratic elution), at a flow rate of 0.6 mL.min^{-1}. Calibration curves and peak integration were performed as already described by Santos et al. [41].

2.3. Determination of Antibacterial Activities

Gram-positive bacteria species used to assess antibacterial activity of fermented whey dilutions were *Listeria monocytogenes* NCTC 11994 (serovar 4b), *L. innocua* ATCC 33090 (type strain) (Instituto Superior de Agronomia—ISA, Lisboa, Portugal), *Staphylococcus aureus* ATCC 6538 (Instituto Nacional de Saúde Dr. Ricardo Jorge—INSA, Lisboa, Portugal), *Bacillus subtilis*, and *B. cereus*, which were isolated from a chickpea salad with parsley at the Food Microbiology Laboratory (FML-INSA). The Gram-negative strains tested were: *Salmonella* enterica serovar Typhimurium ATCC 14028 (ISA), *S.* enterica serovar Goldcoast NCTC 13175, *Escherichia coli* NCTC 9001, *Escherichia coli* O157:H7 NCTC 12900 (verotoxin negative), *Pseudomonas aeruginosa* ATCC 27853, *P. fluorescens* ATCC 13525 (INSA), *Pantoa agglomerans*, and *Citrobacter freundii*, which were isolated from a lettuce salad at FML-INSA. Strains were recovered from a culture at −80 °C, into 10 mL of Brain Heart Infusion Broth (BHI) (Oxoid, Bansingstone, UK) for two consecutive cultures at 24 h intervals, inoculated afterwards on Tryptone Soya Agar (TSA) (bioMérieux® SA, Marcy l'Étoile, France) and incubated at 37 °C ± 1 °C overnight. Chlorine was used as a positive control.

2.3.1. Minimum Inhibitory Concentration (MIC) Throughout the 120 h Fermentation

In a first approach, antibacterial activities were tested throughout the length of the five-day fermentation, with *L. monocytogenes* and *Escherichia coli* O157.H7. Minimum Inhibitory Concentration (MIC) determinations were determined using suspensions of *L. monocytogenes* and *Escherichia coli* O157:H7 from overnight cultures, which were prepared in Müller-Hinton Broth medium ((Biokar Diagnostics™ Beauvais, France) adjusted to a final optical density (OD) of 0.12 (2×10^8 cfu·mL^{-1}) [42], read at 546 nm in a spectrophotometer (Boeco S-20, Hamburg, Germany). Further decimal dilutions were made in the same medium to obtain final suspensions of 2×10^5 cfu·mL^{-1}. MIC values were assessed in sterile 96-well plates (Greiner Bio-one, Germany), using the micro dilution method of Bouhdid et al. [43]. Plates were incubated for 24 h ± 2 h, at 37 °C ± 1 °C, and the OD was read at 546 nm (Synergy HT, Biotek, Winooski, VT, USA) in the beginning of the inoculation and at the end of the assay.

2.3.2. Well-Diffusion Assay

After selecting the day that presented the highest antibacterial activity, fermented whey (100%) was tested on a second screen for its antibacterial activity against several food outburst-related bacterial species, using the well-diffusion assay described by Rizzello et al. [11], modified by incorporation of inoculum into Plate Count Agar (bioMerieux, Marcy l'Etoile, France), 10 mL per plate, overlaid with 10 mL of water-agar (2% *w/v*). Wells of 15 mm were cut into agar plates and 200 μL of 120 h fermented whey were placed in each well. Plates remained at 4 °C for 4 h to allow diffusion.

To test dose-dependence, we also compared non-fermented whey, fermented whey, and milk against the strain of *L. monocytogenes*, using diluted solutions of 100, 50, and 25% (*v/v*). After incubation at 37 °C ± 1 °C, 24 h ± 2 h, plates were measured with a caliper rule for inhibition zone diameters, with averages of three independent trials calculated with standard deviations. As a positive control we used an anionic detergent surfactant and susceptibility was recorded by average diameters > 4 mm [44].

2.4. Impact of Fermented Whey as a MP Lettuce Sanitizer

Loose leaf lettuce (*Lactuca sativa* var. *crispa*) samples were purchased at a local biological market, always from the same organic grower, on the same day of the experiments. Samples were treated in parallel with whey solution and a 110 ppm chlorine solution by dissolving Amokina® (Angelini, Portugal) in sterile water, according to the manufacturer instructions. Outer leaves of the lettuces with signs of damage were discarded; the inner ones were cut using a cylindrical metal cutter of 6 cm diameter in order to take representative portions of all parts of the plant tissue. Circles of lettuce leaves were than washed in distilled water. Subsequently, three different solutions were used to sanitize and wash the lettuce shreds: (1) Distilled water acting as the reference; (2) chlorinated solution (110 ppm of Amokina®), as previously described; and (3) a 75% (*v/v*) fermented whey solution in distilled water [41]. The three different treatments were carried out in plastic bags filled with about 200 g of lettuce shreds, from the three different zones of the lettuces (outer, inner, and middle leaves) immersed in 1 L of the sanitizing or washing solutions. Bags were sealed and soaked for 5 min at 4 °C using an incubator with orbital shaking (Panasonic MIR 154, Gunma, Japan). After that, lettuce shreds were rinsed with sterile distilled water to remove sanitizers, and finally, the excess surface water was removed by a handheld salad spinner (IKEA Tokig, Lisbon, Portugal) for about 30 s. Processed lettuce were pooled and packed at 100 g of shredded lettuce per bag in heat-sealing bags (300 × 230 mm) of 30 μm oriented polypropylene (Amcor Flexibles Neocel, Portugal), graciously granted by Campotec SA. A bag prepared with 100 g of lettuce just soaked in distilled water, whose excess was also removed by the salad spinner, was marked as day 0 and served as the reference. To evaluate O_2 and CO_2 changes, three independent bags were separated. Samples were stored at 4 °C for subsequent evaluation of O_2 and CO_2 changes, pH, texture, color, sensory quality, and microbial growth, on processing day (day 0) and after 1, 3, 5, 7, 9, and 10 days of storage. The experiments were performed in three independent trials.

2.4.1. pH Measurement

The samples were prepared by mixing 10 g of each sample with 20 mL of deionized water in sterile stomacher bags (Seward Limited, London, UK) and then homogenized in a Stomacher (Model 400 Circulator, Seward Limited, London, UK) for 2 min at regular speed. Afterwards, the pH of samples was measured with a Metrohm 827 pH Lab with a 6.0228.010 Primatrode, electrode with integrated temperature sensor (NTC) (Metrohm, Herisau, Switzerland).

2.4.2. Texture Analysis

The texture features of lettuce samples were evaluated in a puncture test with a texturometer TA-XTplus (Stable MicroSystems, Godalming, UK) using a 5 kg load cell. Each sample was fixed between a perforated plate and an acrylic ring with a 9.7 mm hole, as in a sandwich, and kept in place by a plastic clip to avoid any slipping. A 2 mm diameter inox probe penetrated (5 mm distance) the sample through the hole in the rin, at 1 mm/s crosshead speed. From the force-distance texturogram, two parameters were evaluated: firmness, as the maximum rupture force (N) in the yy axis; and brittleness, defined as the rupture deformation (mm), corresponding to the distance of the peak on the xx axis.

2.4.3. Color Measurement

Sample color was measured on the CIELAB L*a*b* chromatic space with a Minolta CR-300 colorimeter (Minolta, Osaka, Japan) with standard illuminant D65 and a visual angle of 2°. Tri-stimulus color coordinates (CIELAB system) were used to measure the degree of lightness (L*) which ranges from 0 (black) to 100 (white), redness (a) ranging from −60 (green) to +60 (red) and yellowness (b) ranging from blue (−60) to yellow (+60). The instrument was initially calibrated using a white ceramic tile (L*97.21; a* 0.13; b* 2.00). The measurement was performed by placing a piece of lettuce directly under the sensor and at least 30 measurements were done by treatment and day of evaluation, covering three different leaves coloration picked by naked eye selection (dark green, median green and light green).

2.4.4. Sensory Evaluation

Evaluation of sensory quality of the samples, after washing with the sanitizing agents, along the ten days of storage, was performed by a panel of ten untrained members. Panelists were required to evaluate changes in visual quality (appearance/freshness), texture, flavor, off-odors, color (browning) and overall acceptability of samples. Samples were scored by an hedonic scale from 1 to 9 with descriptors anchored at both ends (dislike, not characteristic of the product and like very much, very characteristic of the product) to describe attributes considered. The limit of acceptance from the consumer's point of view is set to be 5, and values below this point indicate unacceptable samples.

2.4.5. Microbial Analysis

Enumeration of Aerobic Microorganisms at 30 °C (AM), Psychrotrophic Microorganisms (PM) and Lactic Acid Bacteria (LAB) were performed in duplicate plates as previously described by Santos et al. [41]. Counts were performed for lettuce shreds before and after sanitizing treatments, to monitor microbial development during storage at 1, 3, 5, 7, 9 and 10 days in three independent trials.

2.5. Statistical Analysis

Statistical analysis was made by applying variance analysis, the one factor (ANOVA), and post-hoc multiple comparisons (Tukey test).

3. Results and Discussion

3.1. Longer Fermentation Periods Are Important for the Highest Lactic Acid Production and the Lowest Sugar Content

Although lactic acid production is of importance, the major aim of our research was to obtain a disinfecting agent with immediate application to spin-off as an easy, low-cost, technology. While most studies use short fermentation periods such as 24 h [33,39], in the present work we allowed the fermentation to last up to 120 h. The metabolite profile obtained throughout fermentation period is graphed in Figure 1, which shows the amounts of several metabolites: lactose, galactose, acetic and lactic acids, as well as the pH throughout the 120 h. Results show that although the pH dropped from 6.6 to 3.9 during the first 24 h of fermentation, lactic acid production continued to increase through the 120 h, supporting the need to use longer fermentation times, as opposed to the usual procedure in most studies. Lactic acid was the main acid detected, but acetic acid (0.89 g·L^{-1}) was produced after 120 h. Starting from 30 g·L^{-1} lactose in the unfermented whey, the starter led to a mass conversion rate of lactose into lactic acid of 0.56, while residual lactose was 2.61 g·L^{-1}. The yield of converting lactose into lactic acid was therefore higher in our work when compared to other works, such as Plessas et al. [45] who reported a 0.47 conversion rate. Also, the fact that lactose was reduced to concentrations lower than 3 g·L^{-1} makes it suitable to be used in food products as a disinfectant without the problems of lactose intolerance [46].

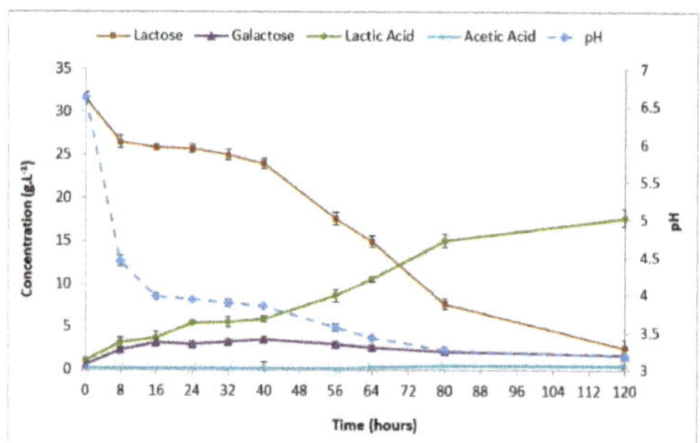

Figure 1. Metabolite profile of whey during fermentation by mesophilic starter bacteria. Results reflect the average of three independent trials ± standard deviation.

Hence, overall results show that our fermentation protocol has the potential to reduce the environmental impact of whey by reducing its organic content.

3.2. Antibacterial Activity Is Also Dependent on the Length of Fermentation

Although several works have already shown that fermented whey can be used as an antibacterial agent in minimally processed salads, being able to reduce overall microorganism growth [8,40], most studies also use fermentation periods around 24 to 72 h [33,39]. Since lactic acid production was higher in longer fermentation periods, we set out to search the best timing along fermentation for higher antibacterial activity. Hence, whey samples were collected daily throughout the 120 h, and firstly tested against reference Gram-positive *Listeria monocytogenes* 4b and Gram-negative *Escherichia coli* O157:H7, and MIC were determined for each day (Table 1). Results show that our fermented whey reduced the growth of both bacteria, from day 1 in *Escherichia coli* O157:H7 and from day 4 in *L. monocytogenes* 4b, concomitantly to the increase in lactic acid produced (Figure 1), respectively, of approximately 5 and 15 g·L^{-1}. However, more importantly, activities were significantly higher on the final day of fermentation, where MICs were more significantly reduced. Thus, our results indicate that fermentation periods longer than 72 h should be preferred in this type of study to obtain the best antibacterial activities.

Table 1. Minimum inhibitory concentrations (MIC) of fermented whey to control two selected bacterial strains *.

Days of Fermentation	Strains	
	Listeria monocytogenes 4b (MIC)	*Escherichia coli* O157:H7 (MIC)
1	NI	25.0
2	NI	25.0
3	NI	25.0
4	25.0	25.0
5	1.56	3.13

Note: * NI = No inhibition observed. Figures are averages of three independent experiments, expressed in % (*v/v*) of whey diluted in growth media.

3.3. Five Days-Fermented Whey Presents Dose-Dependent Antibacterial Activities and Is Effective Against Several Bacterial Species

Based on the results in Figure 1 and Table 1, whey fermented for 120 h was selected to further analyze antibacterial activities towards thirteen pathogenic and spoilage strains, and results are presented in Figure 2. Results show that growth inhibition was induced in all strains, either being Gram-negative or Gram-positive species, corroborating a broad range of antibacterial activity. Subsequently, 120 h whey samples were evaluated against *L. monocytogenes* and assessed at three different dilutions (Table 2) and compared to milk and non-fermented whey. Results show there was a dose-dependent effect in 120 h fermented whey, whereas neither milk nor unfermented whey exhibited an antibacterial effect ($p < 0.05$) (Table 2).

Microorganisms	Ømm		
	Test	+ Control	
E. coli	25.07±0.15	26.17±0.12	*Bacillus_subtitlis*
E. coli O157	25.27±0.06	25.00±0.10	
S. Goldcoast	25.10±0.17	18.10±0.20	*Citrobacter_freundii*
S. Thiphymurium	22.43±0.15	20.23±0.12	
Pantoa agglomerans	31.23±0.21	26.07±0.15	*Pseudomonas fluorecens*
Citrobacter freundii	30.10±0.17	25.20±0.30	
P. aeruginosa	25.10±0.17	23.17±0.25	
P. fluorenscens	33.97±0.21	NI	*Escherichia coli*
L. monocytogenes	31.20±0.17	30.13±0.21	
L. innocua	23.23±0.23	21.97±0.06	
S. aureus	25.10±0.17	30.40±0.26	*Salmonella thiphymurium*
B. cereus	31.47±0.21	35.13±0.06	
B. subtilis	22.10±0.10	17.10±0.30	

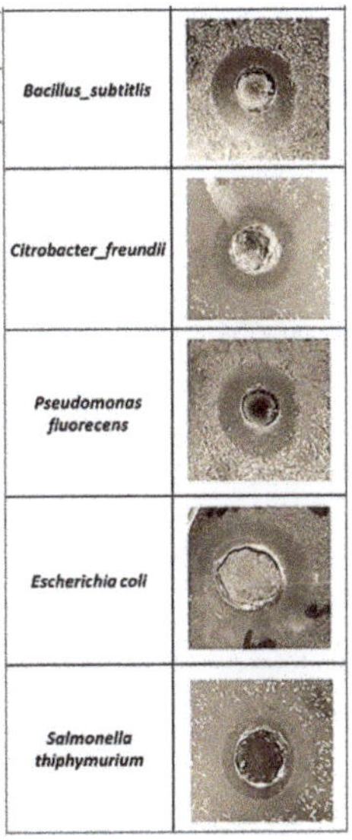

Figure 2. Inhibition halos of different bacteria exposed to 120 h fermented whey. Chlorine was used as a positive control. Results are the average of three replicates ± standard deviation. Note: NI = No inhibition observed.

Table 2. Halos of inhibition of *L. monocytogenes* growth exposed to 120 h fermented whey, milk, or non-fermented whey diluted in water.

	Inhibition Halo (mm)		
Dilutions in Water (*v/v*)	**Milk**	**Non-Fermented Whey**	**Fermented Whey ****
25%	0	0	20.70 ± 0.05 [+]
50%	0	0	24.70 ± 0.05 [*]
100%	0	0	31.20 ± 0.17 [#]

Note: ** Corresponding lactic acid concentrations ($g \cdot L^{-1}$): [+] 4.59; [*] 9.19; [#] 18.38, respectively.

3.4. Industrial Whey Is an Effective Disinfecting Agent When Applied to Lettuce, with Better or Similar Results When Compared Chlorine

Because we aimed to test our 120 h fermented whey in a more realistic manner, we set out to determine its efficacy as a lettuce disinfectant, testing quality indices such as texture and color, including its evaluation by a sensory panel, and also assessing its microbiological quality.

3.4.1. O_2 and CO_2

Table 3 summarizes the monitoring of the O_2 contents and production of CO_2 in the lettuce bags over the 10 days of cold storage. While oxygen was practically kept constant around 19% for the reference samples treated with water, for shredded lettuce sanitized with fermented whey, a reduction from 19 to 18% on the first day was observed, and then it remained practically constant. With chlorine, the reduction continued until the third day to values around 16%.

Table 3. Oxygen and carbon dioxide measurements for packed sanitized shredded lettuce with different sanitizer treatments, over 10 days in cold storage.

Days of Storage	O_2 (%)			CO_2 (%)		
	Water	Chlorine	Whey	Water	Chlorine	Whey
Day 0	19.79 ± 0.41	19.9 ± 0.61	19.4 ± 0.50	0.01 ± 0.04	0.00 ± 0.00	0.02 ± 0.07
Day 1	19.31 ± 0.34	18.5 ± 0.86	18.0 ± 0.71	0.06 ± 0.09	0.43 ± 0.25	0.60 ± 0.32
Day 3	19.44 ± 0.55	16.2 ± 2.31	17.5 ± 1.69	0.23 ± 0.32	0.95 ± 0.78	0.46 ± 0.49
Day 5	19.22 ± 0.71	17.1 ± 1.68	17.7 ± 1.90	0.57 ± 0.46	0.96 ± 0.80	0.56 ± 0.61
Day 7	19.02 ± 0.81	17.5 ± 1.80	17.8 ± 1.42	0.56 ± 0.47	0.93 ± 0.83	0.63 ± 0.58
Day 9	18.97 ± 0.91	16.5 ± 1.86	17.8 ± 2.00	0.64 ± 0.54	1.33 ± 0.70	0.66 ± 1.01
Day 10	19.20 ± 0.94	16.7 ± 1.18	17.3 ± 2.17	0.61 ± 0.57	1.33 ± 0.52	0.78 ± 0.78

For the CO_2 production, more pronounced differences are shown, becoming apparent that the bags treated with chlorine showed a higher reduction in O_2 levels and a larger production of CO_2 compared to the reference and whey treated bags. Consumption of O_2 and production of CO_2 could come from the respiration of the lettuce leaves and AM, as well as further production of CO_2 from anaerobic fermenting microorganisms. Although results suggest that water- and whey-treated samples have lower biological activity, the higher changes detected in the samples treated with chlorine could be due to the response of the lettuce cell leaves to the chemical stress of the chlorinated compounds.

3.4.2. pH

From Figure 3, we can observe that there are no major differences between the pH of the different treatments, with a tendency for a slight increase along the 10-day period. The reference, i.e., lettuce washed with distilled water, is the sample where the increase of pH is significantly ($p < 0.05$ i.e., $p = 8.8 \times 10^{-6}$) higher, from 6.16 to 7.00, but still less than a unit of variation for a storage period of ten days. Martin-Diana et al. [8] reported a similar increase in pH in a study with lettuce and carrots and stated that this was a good index for the potential quality maintenance of the products.

According to Beuchat [47], vegetable products retain adequate quality in a pH range of 5–6.5. Our results show that samples treated with whey exceeded this value (6.57) only on day 9, and the samples treated with chlorine exceeded it on day 7 (6.75).

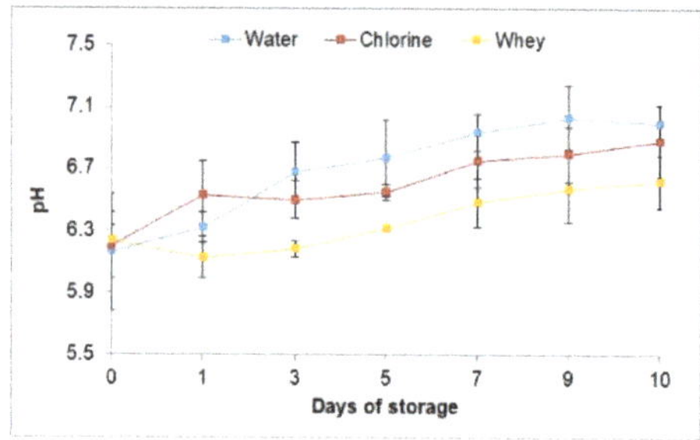

Figure 3. The pH measurements of the lettuce after sanitizer treatments, over 10 days in cold storage. Results represent the average of at least three replicate experiments (n = 3) ± SD.

3.4.3. Texture

Texture is a major quality parameter, and together with color they are crucial for consumer acceptance. Texture measurements evaluated two parameters: resistance to puncture as a force in N, and deformation at rupture in length (mm), generally described as firmness and brittleness (the lower the deformation, the more brittle the material is), respectively. Results are shown in Figure 4a,b. As seen in Figure 4a, the firmness of the lettuce shreds was not substantially affected by any of the treatment and variation within 10 days of storage was quite small. As observed with the pH variation, the reference treatment with water is where the highest difference is found—74 mN, which is significantly ($p = 2.2 \times 10^{-6}$) less firm, and 0.177 mm which is less brittle ($p = 0.0027$) after 10 days in cold storage, i.e., a reduction of 12% in firmness and 7% in brittleness, whereas with whey there was a significant increase in firmness of 69 mN ($p = 8.2 \times 10^{-6}$), corresponding to a 10% increment and no significant variation on brittleness ($p > 0.05$), which is due to the calcium ions present in the whey solution that might reinforce the walls of the lettuce cells.

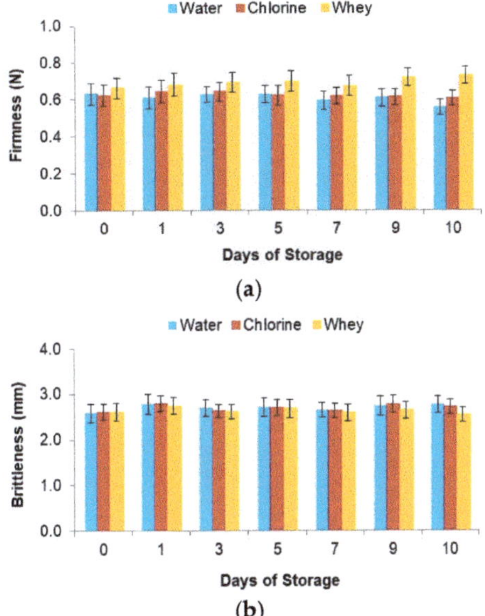

Figure 4. Firmness (**a**) and Brittleness (**b**) of the lettuce after sanitizer treatments, over 10 days in cold storage. Results represent the average of at least three replicate experiments (n = 3) ± SD.

3.4.4. Color Measurement

In Table 4, the lightness L* color parameter for shreds of mid-tone, i.e., median green, can be seen. These results are too scattered to be able to differentiate between sanitizing treatments. The same was observed for the other color coordinates a* and b*. Although 30 shots were taken for each measurement and standard deviations are not high, the chances of taking shreds from different color groups are high and conclusive results cannot be drawn from these experiments. Nevertheless, it can be said that color was not dramatically affected by all treatments.

Table 4. Lightness (L*) color parameter of the lettuce after sanitizer treatments, over 10 days in cold storage. Results represent the average of at least three replicate experiments (n = 3) ± SD.

Days Storage	Water			Chlorine Solution			Fermented Whey Solution		
	L*	a*	b*	L*	a*	b*	L*	a*	b*
0	72.0 ± 4.4	−20.9 ± 1.4	39.0 ± 1.1	97.6 ± 2.4	−19.8 ± 1.2	37.2 ± 0.9	72.1 ± 1.7	−21.3 ± 0.9	38.3 ± 1.7
1	70.2 ± 3.3	−20.8 ± 2.4	33.1 ± 1.2	72.7 ± 3.2	−20.4 ± 0.9	36.7 ± 1.0	103.3 ± 2.8	−21.3 ± 0.6	40.9 ± 2.5
3	71.6 ± 3.7	−17.3 ± 1.4	33.1 ± 1.9	69.2 ± 3.4	−20.7 ± 0.9	39.3 ± 2.5	107.1 ± 3.1	−20.9 ± 0.9	37.8 ± 2.2
5	101.8 ± 2.3	3.8 ± 1.3	2.4 ± 1.8	100.5 ± 1.5	−19.2 ± 0.8	34.3 ± 1.3	61.5 ± 3.3	−19.5 ± 1.3	36.3 ± 2.3
7	74.5 ± 2.9	−19.1 ± 1.4	36.1 ± 1.9	95.2 ± 1.9	−10.8 ± 1.0	17.8 ± 2.3	64.9 ± 2.2	−20.2 ± 0.8	37.4 ± 2.2
9	94.2 ± 3.8	−19.4 ± 1.7	39.7 ± 2.7	98.5 ± 1.8	−20.1 ± 1.5	35.2 ± 2.7	95.0 ± 2.5	−19.6 ± 1.5	37.9 ± 2.3
10	97.1 ± 4.0	−19.0 ± 1.6	36.5 ± 1.7	105.6 ± 3.3	−16.2 ± 1.6	35.8 ± 1.9	93.8 ± 3.3	−19.8 ± 1.6	39.9 ± 2.6

3.4.5. Sensory Evaluation

From the sensory evaluation, one can see the resulting spider diagrams for chlorine and whey treatments in Figures 5a and 5b, respectively. Therefore, from the spider diagrams, one can see that all the attributes were highly scored, with all above the midpoint 5, over the 10 days of cold storage.

To compare the two treatments, overall acceptability was chosen to represent the sensory results, and in Figure 6 it can be seen that there are two major plateaus. The first showed top scores until day 5, and the second with scores around 6 from 7 to the 10th day of cold storage, probably related to the scores obtained for off-odors and flavor. From these results, no differences can be seen for the two treatments, with whey or with chlorine.

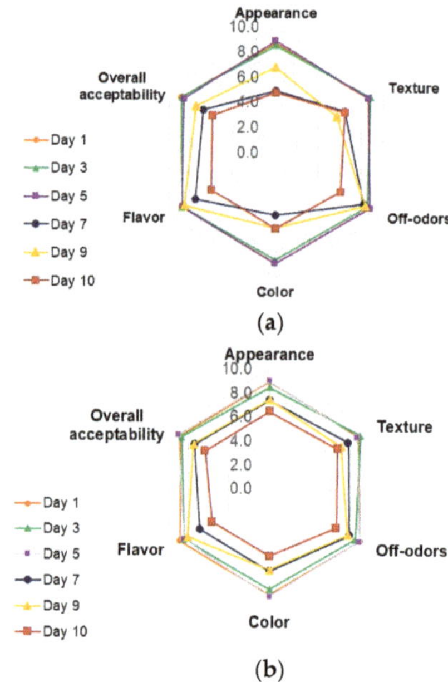

Figure 5. Sensory scores of the lettuce treated with chlorine solution (**a**) and Fermented whey (**b**) over 10 days in cold storage. Results represent the average of at least three replicate experiments (n = 3) ± SD.

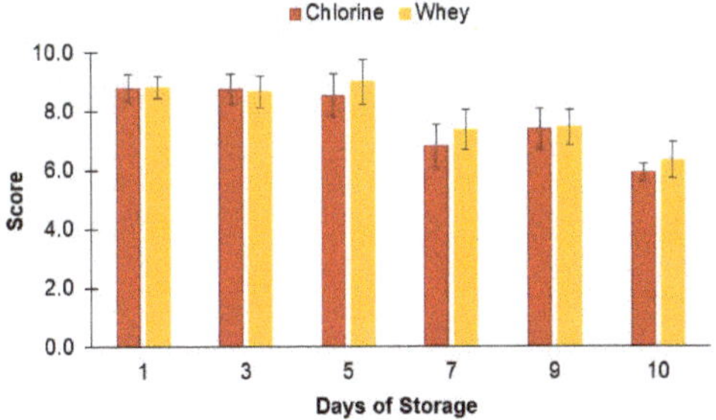

Figure 6. Overall acceptability of the lettuce treated with fermented whey and chlorine over 10 days in cold storage. Results represent the average of at least three replicate experiments (n = 3) ± SD.

3.4.6. Microbial Markers

1. Aerobic Microorganisms at 30 °C

Enumeration of AM (aerobic microorganisms at 30 °C) is an indicator of quality and gives an estimate of total viable populations, both endogenous and contaminant microbiota, since microorganisms are inevitably introduced during manipulations [48]. However, in general, the initial contamination of vegetables reflects the microbiota environment in which they were grown [49,50]. Before treatment (day 0), the initial AM was 6.10 log cfu·g^{-1}. Other works have found similar results in whole vegetables [28,51,52]. As can be observed in Figure 7a, AM counts sharply decreased after all treatments ($p < 0.05$). Throughout the storage, samples treated with whey always presented a significantly lower AM count ($p < 0.001$) when compared to water and chlorine, independent of time ($p < 0.001$). This is a strong indication of the effectiveness of fermented whey as an alternative disinfectant.

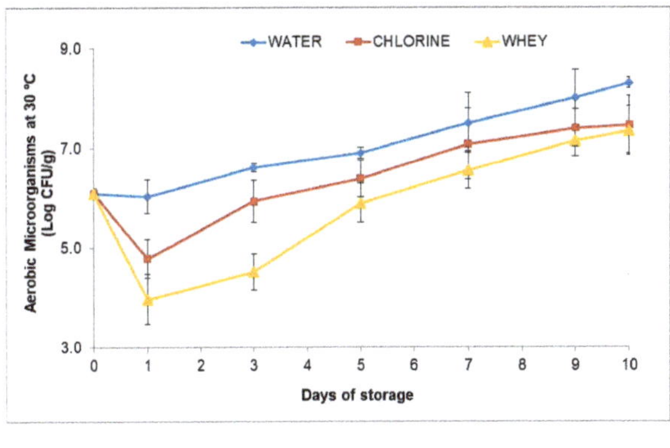

(a)

Figure 7. *Cont.*

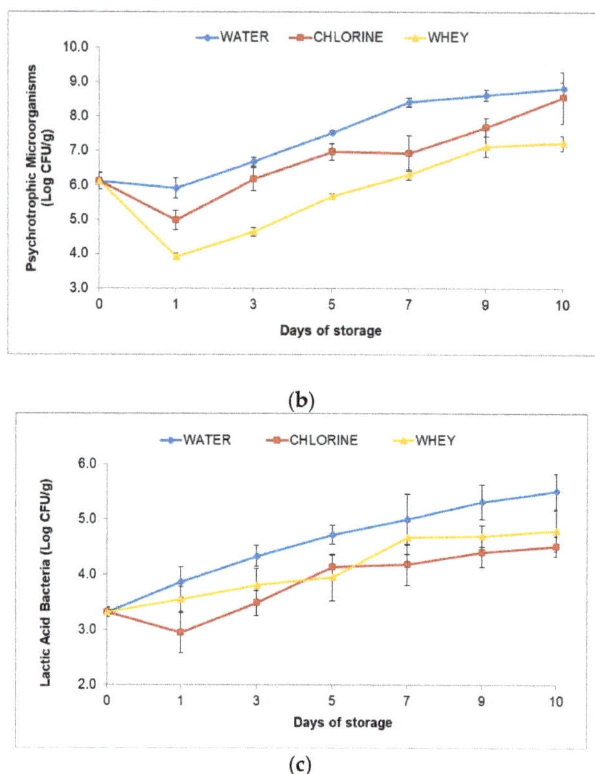

Figure 7. Packed shredded lettuce: effect of sanitation on shredded lettuce Aerobic Microorganisms (**a**), Psychotropic Microorganisms (**b**), and Lactic Acid Bacteria (**c**) over 10 days in cold storage. Results represent the average of at least three replicate experiments (n = 3) ± SD.

2.   Psychrotrophic Microorganisms

Similarly to AM, psychrotrophic microorganisms (PM) counts were significantly reduced ($p < 0.05$) in all treatments at day 1 (Figure 7b). Also, the highest count reduction ($p < 0.001$) to 2.20 log cfu·g^{-1} occurred in samples treated with whey. With chlorine, the count reduction found was 1.13 log cfu·g^{-1}. In the case of the PM, both for the lettuce treated with whey as well as with chlorine, after 5 days in cold storage, there were no significant differences between both treatments in respect to these counts.

3.   Lactic Acid Bacteria

Because whey contains naturally high counts of LAB (in this case 10^7 cfu·g^{-1}), which could induce a higher bacterial load into the product, we investigated the evolution of LAB counts in all treatments throughout cold storage time, and results are presented in Figure 7c. As can be seen, although on day 1 there was a significant difference between LAB counts in all treatments, from day 3 onward, the presence of live LAB from fermented whey did not confer a significant difference ($p > 0.05$) when compared to chlorine and water treatments. This type of bacteria is normally present in MP vegetables in loads from 2 to 6 log cfu·g^{-1} (Figure 7c; at day 0, our reference sample was around 3.5 log cfu·g^{-1}), so LAB counts ended up being similar in all treatments tested, meaning that the introduction of LAB from whey would not pose any noticeable difference to the final product. Furthermore, it has been shown that the presence of these LAB could have a positive antimicrobial effect in lettuce and apple taste [35]. Additionally, the presence of lactic acid, antibacterial peptides, and low pH are contributive factors for the inhibitory effects of whey [53–55].

4. Conclusions

In conclusion, our overall results showed that the established fermentation for industrial whey has the potential to be a low-cost, scalable process to reduce the environmental impact of whey by reducing its organic content, and holds strong potential as an effective disinfecting agent when applied to lettuce, with better or similar results when compared to a 110 ppm chlorine solution. Also importantly, it did not alter the quality parameters of the shredded loose-leaf lettuce, which is highly prone to decay. Color was not substantially affected, and panelists were not able to discriminate from chlorine treatments. Furthermore, lettuce shreds treated with fermented whey showed a slight reinforcement within the cold storage time. Overall, these results not only suggest that fermented whey is indeed as effective as chlorine, but also corroborates that our technology of whey fermentation is effective in maintaining the quality of lettuce throughout storage.

Author Contributions: I.S., M.A.S.S.F., and M.I.S.S. conceived and planned the experiments. I.S., P.F., S.M., M.I.S.S., and A.I.G.L. participated in samples preparation and analysis, data analysis, and interpretation of the results. I.S., P.F., M.I.S.S., and A.I.G.L. wrote the manuscript with input from all authors. I.S., R.M.S.B.F., L.P., and M.A.S.S.F. supervised the research work, contributed to the discussion of the data, and revised the manuscript. All authors contributed suggestions and comments for the final version of the manuscript.

Funding: This research was funded by Fundação para a Ciência e a Tecnologia (Portugal, UID/AGR/04129/2013 (LEAF)).

Acknowledgments: The authors would like to acknowledge Sara Bernardes Silva, Technical Director of Indústria de Laticínios SA, Portugal, for kindly providing the cheese whey used in this work; Rosália Furtado, Maria João Barreira, Anabela Coelho, from the Food Microbiology Laboratory of Instituto Nacional de Saúde Dr. Ricardo Jorge (INSA) for providing some strains used in this work and for technical assistance; Isabel Barroso and Lidia Joaquim from Universidade Lusófona de Humanidades e Tecnologias (ULHT), for technical assistance.

Conflicts of Interest: The authors declare no conflict of interest.

References

1. Ryan, M.P. The biotechnological potential of whey. *Rev. Environ. Sci. Biotechnol.* **2016**, *15*, 479–498. [CrossRef]
2. Prazeres, A.; Carvalho, F.; Rivas, J. Cheese whey management: A review. *J. Environ. Manag.* **2012**, *110*, 48–68. [CrossRef]
3. Carvalho, F.; Prazeres, A.; Rivas, J. Cheese whey wastewater: Characterization and treatment. *Sci. Total Environ.* **2013**, *445*, 385–396. [CrossRef]
4. Murari, C.S.; Moraes, D.C.; Bueno, G.F.; Del Bianchi, V.L. Avaliação da redução na poluição dos laticínios, a partir da fermentação do soro de leite em etanol pela levedura *Kluyveromyces Marxianus* 229. *Revista do Instituto de Laticínios Cândido Tostes, Juiz de Fora* **2013**, *68*, 42–50. [CrossRef]
5. Lievore, P.; Simões, D.R.S.; Silva, K.M.; Drunkler, N.L.; Barana, A.C.; Nogueira, A.; Demiate, I.M. Chemical characterisation and application of acid whey in fermented milk. *J. Food Sci. Technol.* **2015**, *52*, 2083–2092. [CrossRef]
6. Smithers, G. Whey and whey proteins—From "gutter-to-gold". *Int. Dairy J.* **2008**, *18*, 695–704. [CrossRef]
7. Brandelli, A.; Daroit, D.J.; Corrêa, A.P.F. Whey as a source of peptides with remarkable biological activities. *Food Res. Int.* **2015**, *73*, 149–161. [CrossRef]
8. Martin-Diana, A.; Rico, D.; Frias, J.; Mulcahy, J. Whey permeate as a bio-preservative for shelf life maintenance. *Innov. Food Sci. Emerg.* **2006**, *7*, 112–123. [CrossRef]
9. Panesar, P.; Kennedy, J.; Gandhi, D.; Bunko, K. Bioutilisation of whey for lactic acid production. *Food Chem.* **2007**, *105*, 1–14. [CrossRef]
10. Rico, D.; Martín-Diana, A.; Barat, J.; Barry-Ryan, C. Extending and measuring the quality of fresh-cut fruit and vegetables: A review. *Trends Food Sci. Technol.* **2007**, *18*, 373–386. [CrossRef]
11. Rizzello, C.G.; Losito, I.; Gobbetti, M.; Carbonara, T.; Bari, M.D.; Zambonin, P.G. Antibacterial Activities of Peptides from the Water-Soluble Extracts of Italian Cheese Varieties. *J. Dairy Sci.* **2005**, *88*, 2348–2360. [CrossRef]

12. Almaas, H.; Eriksen, E.; Sekse, C.; Comi, I.; Flengsrud, R.; Holm, H.; Jensen, E.; Jacobsen, M.; Laangsrud, T.; Vegarud, G.E. Antibacterial peptides derived from caprine whey proteins, by digestion with human gastrointestinal juice. *Br. J. Nutr.* **2011**, *106*, 896–905. [CrossRef]

13. Medeiros, G.V.; Queiroga, R.E.; Costa, W.A.; Gadelha, C.A.; Lacerda, R.R.; Lacerda, J.G.; Pinto, L.S.; Braganhol, E.; Teixeira, F.C.; Barbosa, P.S.; et al. Proteomic of goat milk whey and its bacteriostatic and antitumour potential. *Int. J. Biol. Macromol.* **2018**, *113*, 116–123. [CrossRef]

14. Hafeez, Z.; Cakir-Kiefer, C.; Roux, E.; Perrin, C.; Miclo, L.; Dary-Mourot, A. Strategies of producing bioactive peptides from milk proteins to functionalize fermented milk products. *Food Res. Int.* **2014**, *63*, 71–80. [CrossRef]

15. Théolier, J.; Hammami, R.; Labelle, P.; Fliss, I.; Jean, J. Isolation and identification of antimicrobial peptides derived by peptic cleavage of whey protein isolate. *J. Funct. Foods* **2013**, *5*, 706–714. [CrossRef]

16. Dullius, A.; Goettertb, M.I.; Souza, C.V. Whey protein hydrolysates as a source of bioactive peptides for functional foods—Biotechnological facilitation of industrial scale-up. *J. Funct. Foods* **2018**, *42*, 58–74. [CrossRef]

17. Murray, K.; Fan Wu, F.; John Shi, J.; Sophia Jun Xue, S.J.; Warriner, K. Challenges in the microbiological food safety of fresh produce: Limitations of post-harvest washing and the need for alternative interventions. *Food Qual. Saf.* **2017**, *1*, 289–301. [CrossRef]

18. Callejón, R.M.; Rodríguez-Naranjo, M.I.; Ubeda, C.; Hornedo-Ortega, R.; Garcia-Parrilla, M.C.; Troncoso, A.M. Reported foodborne outbreaks due to fresh produce in the United States and European Union: Trends and causes. *Foodborne Pathog. Dis.* **2015**, *12*, 32–38. [CrossRef]

19. Charlier, C.; Goffinet, F.; Azria, A.; Leclercq, A.; Lecuit, M. Inadequate management of pregnancy-associated listeriosis: Lessons from four case reports. *Clin. Microbiol. Infect.* **2014**, *20*, 246–249. [CrossRef]

20. Francis, G.A.; Thomas, C.; O'Breirne, D. Isolation and pulsed-field gel electrophoresis typing of *Listeria monocytogenes* from modified atmosphere packaged fresh-cut vegetables collected in Ireland. *J. Food Protect.* **2006**, *69*, 2524–2558. [CrossRef]

21. Gombas, D.; Chen, Y.; Clavero, R.; Scott, V. Survey of *Listeria monocytogenes* in Ready-to-Eat Foods. *J. Food Protect.* **2003**, *66*, 559–569. [CrossRef]

22. Lianou, A.; Sofos, J.N. A review of the incidence and transmission of *Listeria monocytogenes* in ready-to-eat products in retail and food service environments. *J. Food Protect.* **2007**, *70*, 2172–2198. [CrossRef]

23. World Health Organization & Food and Agriculture Organization of the United Nations. Microbiological Hazards in Fresh Leafy Vegetables and Herbs: Meeting Report. World Health Organization. 2008. Available online: https://apps.who.int/iris/handle/10665/44031 (accessed on 22 November 2018).

24. Scientific Committee on Food. Risk Profile on the Microbiology Contamination of Fruits and Vegetables Eaten Raw. Report of the Scientific Commission on Food 2002, European Commission on Health and Consumer Protection Directorate-General. Available online: https://ec.europa.eu/food/sites/food/files/safety/docs/sci-com_scf_out125_en.pdf (accessed on 22 November 2018).

25. Meireles, A.; Giaouris, E.; Simões, M. Alternative disinfection methods to chlorine for use in the fresh-cut industry. *Food Res. Int.* **2016**, *82*, 71–85. [CrossRef]

26. Francis, G.A.; Gallone, A.; Nychas, G.J.; Sofos, J.N.; Colelli, G.; Amodio, M.L.; Spano, G. Factors affecting quality and safety of fresh-cut produce. *Crit. Rev. Food Sci. Nutr.* **2012**, *52*, 595–610. [CrossRef]

27. Warriner, K.; Namvar, A. Recent advances in fresh produce post-harvest decontamination technologies to enhance microbiological safety. *Stewart Postharvest Rev.* **2013**, *9*, 1–8. [CrossRef]

28. Santos, M.I.; Cavaco, A.; Gouveia, J.; Novais, M.R.; Nogueira, P.J.; Pedroso, L.; Ferreira, M.A.S.S. Evaluation of minimally processed salads commercialized in Portugal. *Food Control* **2012**, *23*, 275–281. [CrossRef]

29. Gil, M.I.; Selma, M.V.; López-Gálvez, F.; Allende, A. Fresh-cut product sanitation and wash water disinfection: Problems and solutions. *Int. J. Food Microbiol.* **2009**, *134*, 37–45. [CrossRef]

30. Akbas, M.Y.; Ölmez, H. Inactivation of *Escherichia coli* and *Listeria monocytogenes* on iceberg lettuce by dip wash treatments with organic acids. *Lett. Appl. Microbiol.* **2007**, *44*, 619–624. [CrossRef]

31. Beuchat, L.R.; Adler, B.B.; Lang, M.M. Efficacy of chlorine and a peroxyacetic acid sanitizer in killing *Listeria monocytogenes* on iceberg and romaine lettuce using simulated commercial processing conditions. *J. Food Protect.* **2004**, *67*, 1238–1242. [CrossRef]

32. Tian, J.; Ban, X.; Zeng, H.; Huang, B.; He, J.; Wang, Y. In vitro and in vivo activity of essential oil from dill (*Anethum graveolens* L.) against fungal spoilage of cherry tomatoes. *Food Control* **2011**, *22*, 1992–1999. [CrossRef]

33. Guha, A.; Banerjee, S.; Bera, D. Production of lactic acid from sweet meat industry waste by *Lactobacillus delbruki*. *Int. J. Res. Eng. Technol.* **2013**, *2*, 630–634. [CrossRef]

34. Siroli, L.; Patrignani, F.; Diana, I.; Serrazanetti, D.I.; Gardini, F.; Lanciotti, R. Innovative strategies based on the use of bio-control agents to improve the safety, shelf-life and quality of minimally processed fruits and vegetables. *Trends Food Sci. Technol.* **2015**, *46*, 302–310. [CrossRef]

35. Trias, R.; Bañeras, L.; Badosa, E.; Montesinos, E. Bioprotection of Golden Delicious apples and Iceberg lettuce against foodborne bacterial pathogens by lactic acid bacteria. *Int. J. Food Microbiol.* **2008**, *123*, 50–60. [CrossRef] [PubMed]

36. Bjorkroth, J.; Koort, J. Lactic acid bacteria: Taxonomy and Biodiversity. In *Encyclopedia of Dairy Sciences*, 2nd ed.; Fuquay, J.W., Fox, P.F., McSweeney, P.L.H., Eds.; Academic Press: London, UK, 2011; Volume 3, pp. 45–48.

37. Madureira, A.R.; Soares, J.C.; Amorim, M.; Tavares, T.; Gomes, A.M.; Pintado, M.M.; Malcata, F.X. Bioactivity of probiotic whey cheese: Characterization of the content of peptides and organic acids. *J. Sci. Food Agric.* **2013**, *93*, 1458–1465. [CrossRef] [PubMed]

38. Zacharof, M.; Lovitt, R. Bacteriocins produced by lactic acid bacteria a review article. *APCBEE Proc.* **2012**, *2*, 50–56. [CrossRef]

39. Panesar, P.S.; Kumari, S.; Panesar, R. Potential applications of immobilized β-galactosidase in food processing industries. *Enzyme Res.* **2010**, *2010*, 473137. [CrossRef] [PubMed]

40. Santos, M.I.S.; Lima, A.I.; Monteiro, S.A.V.S.; Ferreira, R.M.S.B.; Pedroso, L.; Sousa, I.; Ferreira, M.A.S.S. Preliminary study on the effect of fermented cheese whey on *Listeria monocytogenes*, *Escherichia coli* O157:H7 and *Salmonella* Goldcoast populations inoculated onto fresh organic lettuce. *Foodborne Pathog. Dis.* **2016**, *13*, 423–427. [CrossRef]

41. Santos, M.I.S.; Martins, S.R.; Pedroso, L.; Sousa, I.; Ferreira, M.A.S.S. Potential bio-activity of whey fermented extract as sanitizer of organic grown lettuce. *Food Control* **2015**, *50*, 477–481. [CrossRef]

42. Gottlieb, C.T.; Thomsen, L.E.; Ingmer, H. Antimicrobial peptides effectively kill a broad spectrum of *Listeria monocytogenes* and *Staphylococcus aureus* strains independently of origin, sub-type, or virulence factor expression. *BMC Microbiol.* **2008**, *8*, 205. [CrossRef]

43. Bouhdid, S.; Abrini, J.; Amensour, M.; Zhiri, A.; Espuny, M.J.; Manresa, A. Functional and ultrastructural changes in *Pseudomonas aeruginosa* and *Staphylococcus aureus* cells induced by *Cinnamomum verum* essential oil. *J. Appl. Microbiol.* **2010**, *109*, 1139–1149. [CrossRef]

44. Pintado, C.M.B.S.; Ferreira, M.A.S.S.; Sousa, I. Properties of whey-protein based films containing organic acids and nisin to control *Listeria monocytogenes*. *J. Food Prot.* **2009**, *72*, 1891–1896. [CrossRef] [PubMed]

45. Plessas, S.; Bosnea, L.; Psarianos, C.; Koutinas, A.; Marchant, R.; Banat, I.M. Lactic acid production by mixed cultures of *Kluyveromyces marxianus*, *Lactobacillus delbrueckii* ssp. *bulgaricus* and *Lactobacillus helveticus*. *Bioresour. Technol.* **2008**, *99*, 5951–5955. [CrossRef] [PubMed]

46. European Food Safety Authority. Scientific opinion on lactose thresholds in lactose intolerance and galactosaemia. *EFSA J.* **2010**, *8*, 1777. [CrossRef]

47. Beuchat, L.R. Surface disinfection of raw produce. *Dairy Food Environ. Sanit.* **1992**, *12*, 6–9.

48. Health Protection Agency. *Guidelines for Assessing the Microbiological Safety of Ready-to-Eat Foods*; Health Protection Agency: London, UK, 2009.

49. Barth, M.; Hankinson, T.; Zhuang, H.; Breidt, F. Microbiological Spoilage of Fruit and Vegetables. In *Compendium of the Microbiological Spoilage of Foods and Beverages*; Sperber, W.H., Doyle, M.P., Eds.; Springer: New York, NY, USA, 2009; pp. 135–183.

50. Tauxe, R.; Kruse, H.; Hedberg, C.; Potter, M.; Madden, J.; Wachsmuth, K. Microbiological Hazards and Emerging Issues Associated with Produce: A Preliminary Report to the National Advisory Committee on Microbiological Criteria for Foods. *J. Food Prot.* **1997**, *60*, 1400–1408. [CrossRef] [PubMed]

51. Abadias, M.; Usall, J.; Anguera, M.; Solsona, C.; Viñas, I. Microbiological quality of fresh, minimally-processed fruit and vegetables, and sprouts from retail establishments. *Int. J. Food Microbiol.* **2008**, *123*, 121–129. [CrossRef] [PubMed]

52. Ponce, A.G.; Roura, S.I.; Del Valle, C.E.; Fritz, R. Characterization of native microbial population of Swiss Chard (*Beta vulgaris, type cicla*). *LWT Food Sci. Technol.* **2002**, *35*, 331–337. [CrossRef]
53. Beuchat, L.R. Use of Sanitizers in Raw Fruit and Vegetable Processing. In *Minimally Processed Fruits and Vegetables*; Alzamora, S.M., Tapia, M.S., Lopez-Malo, A., Eds.; An Aspen Publication: Gaithersburg, MD, USA, 2000; pp. 63–78.
54. Nykänen, A.; Lapveteläinen, A.; Kallio, H.; Salminen, S. Effects of whey, whey-derived lactic acid and sodium lactate on the surface microbial counts of rainbow trout packed in vacuum pouches. *LWT Food Sci. Technol.* **1998**, *31*, 361–365. [CrossRef]
55. Yang, E.; Fan, L.; Jiang, Y.; Doucette, C.; Fillmore, S. Antimicrobial activity of bacteriocin-producing lactic acid bacteria isolated from cheeses and yogurts. *AMB Express* **2012**, *2*, 48. [CrossRef]

Article

Increased Anti-Inflammatory Effects on LPS-Induced Microglia Cells by *Spirulina maxima* Extract from Ultrasonic Process

Woon Yong Choi [1], Jae-Hun Sim [2], Jung-Youl Lee [2], Do Hyung Kang [3] and Hyeon Yong Lee [4,*]

[1]   Department of Medical Biomaterials Engineering, Kangwon National University, Chuncheon 200-701, Korea; cwy1012@hanmail.net
[2]   R&BD Center, Korea Yakult Co. Ltd., Seoul 17086, Korea; jhsim@re.yakult.co.kr (J.-H.S.); jlleesk@re.yakult.co.kr (J.-Y.L.)
[3]   The Jeju International Marine Science Center for Research & Education, Korea Institute of Ocean Science and Technology (KIOST), Ansan 63349, Korea; dohkang@kiost.ac.kr
[4]   Department of Food Science and Engineering, Seowon University, Cheongju 361-742, Korea
*   Correspondence: hyeonl@seowon.ac.kr; Tel.: +82-4-3299-8471

Received: 5 April 2019; Accepted: 22 May 2019; Published: 26 May 2019

Abstract: The *Spirulina maxima* exact from a non-thermal ultrasonic process (UE) contains 17.5 mg/g of total chlorophyll, compared to 6.24 mg/g of chlorophyll derived from the conventional 70% ethanol extraction at 80 °C for 12 h (EE). The UE also showed relatively low cytotoxicity against murine microglial cells (BV-2) and inhibited the production of the inflammatory mediators, NO and PGE_2. The UE also effectively suppresses both mRNA expression and the production of pro-inflammatory cytokines, such as TNF-α, IL-6 and IL-1β, in a concentration-dependent manner. Notably, TNF-α gene and protein production were most strongly down-regulated, while IL-6 was the least affected by all ranges of treatment concentrations. This work first demonstrated a quantitative correlation between mRNA expression and the production of cytokines, showing that suppression of TNF-α gene expression was most significantly correlated with its secretion. These results clearly proved that the anti-inflammatory effects of *Spirulina* extract from a nonthermal ultrasonic process, which yielded high concentrations of intact forms of chlorophylls, were increased two-fold compared to those of conventional extracts processed at high temperature.

Keywords: *Spirulina maxima*; anti-inflammatory effects; neuroproective activities; ultrasonic extraction process

---

1. Introduction

Recently, studies on the discovery and treatment of new diseases have been accelerated by the dramatic development of medical technologies. As our understanding of the human brain has widened, studies on the causes and treatments of diseases related to brain diseases have been intensively conducted [1–3]. Studies on the treatment and prevention of dementia, which is among the most threatening diseases to humans both socially and medically and seriously deteriorates patients' quality of life, have been intensively conducted, substantially advancing our mechanistic understanding of dementia and the development of diverse methods and drugs to treat dementia [3–5]. Specifically, many studies on cognitive function deterioration and preventive drugs have widely accepted the theory that a major cause of cognitive dysfunction is brain nerve cell death, due to persistent exogenous stress or accumulation of oxidants, which lead to cognitive function deterioration and consequent brain diseases, such as dementia [6–8]. Therefore, many studies on cognitive and memory impairments, such as dementia and Parkinson's disease, have concentrated on mechanisms that protect brain nerve cells by preventing the accumulation of metabolic oxidants through antioxidant activity or rapid decomposition of gene oxidants and the development of drugs for such mechanisms [9–11]. Therefore, drugs, such as donepezil, rivastigmine, and galantamine have been developed to control and/or enhance the neuronal signaling pathway linked to p-ERK/p-CREB/BDNF. However, after extended use, these synthetic drugs show adverse side effects, such as vomiting, hand tremors, and heart disease [12–14]. Accordingly, remarkable findings have been reported from studies on natural products that protect cranial nerves by exerting antioxidant activity, and these products have relatively fewer side effects [15–17].

Most of these studies have fundamentally focused on suppressing cognitive function and memory loss through antioxidant-mediated cranial nerve protection. However, rather than exhibiting intense antioxidative effects in the brain nerve cells and their activation sites, natural products suppress the deterioration of cognitive function resulting from overall antioxidative activity throughout the body, so the actual effects have been remarkably lower than expected in most cases [18,19]. Among the recent studies on suppressing cognitive function and memory loss, those featuring natural products with functions beyond simple antioxidant activity to directly suppress memory deterioration by inhibiting inflammation in brain nerve cells have attracted substantial attention [20]. In particular, the brain damage-induced activation of pro-inflammatory cytokines, such as TNF-α and IL-6, can evoke brain inflammation, which results in various neurodegenerative diseases like ischemia, trauma, infection, Alzheimer's disease and Parkinson's disease [20–22]. There is recent strong evidence that continuous stress and diverse inflammation in the brain could be a major and direct cause of dementia and memory impairment, due to the continuous accumulation of inflammatory cytokines in the brain. While many efforts have been made to reduce the production of these inflammatory cytokines to indirectly attenuate degenerative brain diseases, more detailed studies on the mechanism underlying delayed neurodegeneration should be carried out based on our understanding of the blood-brain barrier [23,24]. Therefore, the development of functional natural materials that more efficiently prevent and treat dementia is possible if the materials can selectively inhibit brain nerve cell inflammation and exert antioxidative activity [25,26].

To this end, a variety of natural resources have been developed recently, including *Spirulina* which has attracted much attention. *Spirulina* has been consumed by humans for thousands of years without side effects and has been selected as space food by the US National Aeronautics and Space Administration (NASA). Since *Spirulina* has diverse active substances, such as C-phycocyanin, beta-carotene, chlorophyll, and functional fatty acids, it has been reported to have excellent anti-cancer, immunity enhancing, skin improving, and anti-inflammatory effects and to be useful in treating hypertension, diabetes, and diseases related to metabolism, such as liver dysfunctions [27–29]. In particular, *Spirulina* contains at least 50–100% more beta-carotene and chlorophyll, which are highly antioxidative, than other natural products [27,30]. Therefore, many study findings have been reported promising outcomes of *Spirulina* on brain nerve cell protection from these strong antioxidants, and many positive effects of *Spirulina* in cranial nerve protection are associated with the strong antioxidant activities of chlorophylls [31,32]. Interestingly, *Spirulina* is also known to contain at least 70–90% pure chlorophyll a, which has relatively higher activity than the other subtypes chlorophyll b and c which normally exist in a mixture in most plants. Therefore, studies using the antioxidant activity of *Spirulina* are expected to have great potential [33,34]. However, most of these studies have focused on the antioxidant activity and regulation of antioxidative signaling pathways in the body [32,35,36], while studies on enhancing memory by suppressing brain nerve cell inflammation via chlorophyll derived from *Spirulina* are extremely rare.

Although studies on brain cell anti-inflammation using chlorophylls from *Spirulina* are promising, obtaining a high concentration of chlorophylls from *Spirulina* using existing conventional extraction processes is difficult because chlorophylls are extremely vulnerable to heat and fat solubility [37,38]. Low-temperature extraction is essential to obtain high-concentration chlorophyll that maintains its activity. However, it is difficult to obtain chlorophyll at a sufficient concentration by ordinary low-temperature extraction processes, due to very low extraction yields and longer process time [37,38].

To overcome this difficulty, an ultrasonic low-temperature extraction process will be an excellent alternative to treat heat-sensitive *Spirulina* extract. The ultrasonic process (UE) is a typical low-temperature extraction process that does not apply heat and can efficiently extract heat-sensitive active components through efficient destruction of the cell walls using air cavities generated by ultrasonic vibration even at room temperature or temperatures below 40 °C [39,40]. Therefore, this study investigated the anti-inflammatory mechanism of brain nerve cells associated with the antioxidant activity of chlorophylls in the *Spirulina* extract obtained by low-temperature ultrasonic treatment.

2. Materials and Methods

2.1. Preparation of the S. maxima Extracts

100 g of dried *S. maxima* (cultivated in Korea Institute of Ocean Science and Technology, Jeju Center, Korea) was pretreated with 70% ethanol (*w/w*) at a ratio of 1:10 (*w/w*) for 8 h using an ultrasonicator (AUG-R3-900, ASIA ULTRASONIC, Gyeonggi-do, Korea) installed with a temperature controller at 40 kHz with 800 W of input power and constant room temperature. Then, the pretreated *S. maxima* was further extracted by a reflux condenser (TL-6Point, Misung Scientific Co., Yangjoo, Korea) at 65 °C for 4 h. For the conventional ethanol extraction, 100 g of dried *S. maxima* were extracted with 1 L of 70% ethanol (*w/w*) at 80 °C for 12 h by a reflux condensing extractor. Then, the extracts from each experiment were concentrated using a rotary vacuum evaporator (N-N series, EYELA, Rikakikai Co., Tokyo, Japan) and freeze-dried with a freeze-dyer (PVTFA 10AT, ILSIN, Suwon, Korea) before the use. The extract from each process was expressed as an ultrasonic extract (UE) from the ultrasonic pretreated process at room temperature and ethanol extract (EE) from conventional hot ethanol extraction process at 80 °C, respectively.

2.2. Measurement of the Chlorophyll Contents in the Extracts

The amounts of total chlorophylls in the extracts were estimated by the following method [41]: First, 1.4 g of sodium oxide was dissolved in 40 mL of distilled water and 16.6 g of pyridine was added to make 100 mL of the final volume. Then, 10 mg of the extract was placed in a foil-covered test tube and added with 10 mL of distilled water, and further incubated for 30 min. Thereafter, 2 mL of the suspension was mixed with 5 mL of an alkaline pyridine solution and incubated at 60 °C with aluminum foiled cover, and then allowed to stand for 15 min. After that, the solution was centrifuged at 3000× *g* for 3 min. and the supernatant was transferred to a 10 mL brown flask and stored in a cool place. Then, the absorbance of the sample solution was compared with that of alkaline pyridine solution by a UV spectrophotometer (UTX-20M, Biostep co., Burkhardsdorf, Germany) at 419 nm and 454 nm of wavelengths. The actual concentrations in the extracts were estimated by the following Equation (1):

$$\text{Total chlorophyll (mg/100 g)} = C/S \times 100. \tag{1}$$

C: Chlorophyll (mg/L) = $8.970 \times (7.19 \times A419 \text{ nm} + 3.33 \times A454 \text{ nm})$
S: Weight (mg) of the extracted sample in 2 mL of alkaline pyridine solution
A419 nm: Absorbance at a wavelength of 419 nm
A454 nm: Absorbance at a wavelength of 454 nm

2.3. Measurement of Cell Cytotoxicity and the Production of Nitric Oxide (NO) and Prostaglandin E2 (PGE2)

The cytotoxicity of the *Spirulina* extracts to mouse microglial cells (BV-2, ATCC, Manassas, VA, USA) was observed with the following method [42]. First, BV-2 cells were seeded in a 24-well plate at a concentration of 1×10^5 viable cells/mL with Dulbecco's modified Eagle's Medium (DMEM, Gibco, Carlsbad, CA, USA) enriched with 10% fetal bovine serum (FBS, Gibco, Carlsbad, CA, USA), and 100 U/mL penicillin and 100 µg/mL streptomycin at 37 °C with 5% CO_2. Then, the cells were treated with various concentrations of two samples (UE and EE) with or without 1 µg/mL of lipopolysaccharide (LPS, L2630, Sigma, St. Louis, MO, USA), as well as with no treatment as a control, and cultured for one more day [43]. Next, the culture medium was removed and 1 mg/mL 3-(4,5-methylthiazol-2-yl)-2,5-diphenyltetrazolium bromide (MTT, Sigma) solution was added. The cells were cultured again for 4 h at 37 °C with 5% CO_2 with minimal light. The MTT solution was then removed, 200 µl of dimethyl sulfoxide (DMSO, Sigma) was added to the wells, and the wells were incubated for 30 min in the dark. After that, the absorbance was measured at a wavelength of 570 nm by a microplate ELISA reader (Thermo Fisher Scientific, Waltham, MA, USA). The cytotoxicity was estimated as the ratio of the absorbance of the untreated group (Au) to that of the sample group (As), as shown in the following Equation (2) [44]:

$$\text{Cytotoxicity (\%)} = (1 - \text{Au/As}) \times 100. \tag{2}$$

The concentrations of NO and PGE2 secreted from BV-2 cells were measured with the following method. To measure NO production from BV-2 cells, the BV-2 cells were seeded into 96-well plates at a concentration of 1.0×10^5 cells/mL, certain concentrations of each of the samples were added to the wells, and the cells were incubated for 4 h. Then 1 µg/mL of LPS was added, and the cells were cultured for 24 h at 37 °C in a CO_2 incubator (CB150, Binder, Bohemia, NY, USA) with 5% CO_2. Next, 50 µL of Griess reagent and 50 µL of cell culture supernatant were mixed, added to the cells, and allowed to react for 5 min in 96-well plates. The amount of NO in the medium was measured at 540 nm using an ELISA Reader (Thermo Fisher Scientific, Waltham, MA, USA) with $NaNO_2$ as a standard [45]. To measure PGE2, a PGE2 EIA kit (R&D Systems, Minneapolis, MN, USA) was used by adding the supernatants from BV-2 cells treated with the same procedures described above for NO experiments. The resulting culture media were used to measure the amounts of PGE2 using an ELISA Reader (Thermo Fisher Scientific, Waltham, MA, USA) according to the manufacturer's protocols [46].

2.4. Reverse Transcription Polymerase Chain Reaction (RT-PCR)

To measure the expression of tumor necrosis factor-α (TNF-α), Interleukin-6 (IL-6) and Interleukin-1β (IL-1β) in murine microglial cells, 1×10^5 cells/mL of BV-2 cells were cultured with or without varying concentrations of the *Spirulina* along with 1 µg/mL of LPS, following the same procedures in the NO and PGE_2 measurement experiments. After incubation, RNA was isolated from BV-2 cells using the RNeasy mini kit (Qiagen, Hilden, Germany) according to the manufacturer's protocols. Then 1 µg of total RNA was synthesized into cDNA by a cDNA synthesis kit (RevertAid First Strand Kit, Fermentas, ON, Canada) with incubations at 25 °C for 5 min., 42 °C for 60 min., and finally 85 °C for 5 min. Then, 40 µl of cDNA was mixed with 0.5 µl of each target gene primer for mouse TNF- α, IL-6, IL-1β and β-actin. Forward and reverse primers were provided by the manufacture (Bioneer, Inc., Seoul, Korea) as premade primers as follows: TNF- α (N-4015, 300 bp), IL-6 (N-4013, 155 bp), IL-1β (N4009, 291 bp) and β-actin (N4021, 395 bp), respectively. The mixture was amplified with an RNA PCR kit (Takara, Shiga, Japan) at 94 °C for 30 s, 55 °C for 30 s and 72 °C for 1 min for 35 cycles using a PCR equipment (XP Thermal Cycler, TC-XP, BIOER Tech. Co., Hangzhou, China). Then, the PCR products were analyzed on ethidium bromide (EB, Sigma)-stained 1.2% agarose gels (BioRad co., Berkeley, CA, USA) through electrophoresis with 1-5 V/cm of the applied voltage. The amount of mRNA corresponding to each gene was quantified by Image Processing Analysis in Java (Image J, National Institute of Mental Health, Bethesda, MD, USA) by normalizing to the housekeeping gene β-actin; the values were expressed as the relative amounts (%).

2.5. Measurement of the Secretioin of Pro-Inflammatory Cytokines from BV-2 Cells

The amounts of three different Cytokines, tumor necrosis factor-α (TNF-α), Interleukin 6 (IL-6), and Interleutin-1β (IL-1β) secreted from mouse microglial cells were measured with TNF-α, IL-6 and IL-1β ELISA kits (R&D Systems) as follows [47]. First, BV-2 cells (1×10^5 cells/mL) were inoculated into a 96-well plate and cultured for one day at 37 °C with 5% CO_2. Then, various concentrations of the *Spirulina* extracts and 1 µM of vitamin D3 (cholecalciferol, Sigma) dissolved in 0.1% ethanol were added to 1 µg/mL of LPS for one additional day of culturing. Following the ELISA kit manuals, 50 µL of the assay diluent was added to the wells, and the standard samples were also added to the wells. Then the plate was shaken for one week and left unattended for 2 h at room temperature in the dark. Next, the plate was washed with a wash buffer, 100 µL of the substrate solution was added to the plate, and the plate was incubated for 30 min at room temperature in the dark. After blocking the reaction with a stop solution, the concentrations of the cytokines were measured at 450 nm with an ELISA reader (Thermo Fisher Scientific).

2.6. Statistical Analysis

All experimental data were repeated three times, and statistical significance was analyzed by one-way Analysis of Variance (ANOVA) using the Statistical Analysis System program (SAS, Cary, NC, USA). The difference between the significance levels was set to $p < 0.05$.

3. Results and Discussion

3.1. Total Chlorophyll Concentrations in the Extracts from Two Different Extraction Processes

Table 1 shows the comparison of the total chlorophyll in the extracts and the difference in extraction yields between the conventional 70% ethanol at 80 °C for 12 h extraction and the UE extraction. The amount of chlorophyll in the UE extract was 17.56 mg/g, which is at least three-fold higher than the concentration obtained through extraction with 70% ethanol at 80 °C which was 6.24 mg/g. This result clearly proved that most of the chlorophyll in the extract was destroyed at temperatures exceeding 60 °C resulting in extremely low chlorophyll content. Therefore, this result reconfirms that heat-sensitive substances, 'such as chlorophyll can be extracted' in their intact forms with the least loss of their activities only when they are extracted at low temperatures associated with UE. Additionally, low-temperature extraction does not affect the activity of the target substance, and the extraction yield is reduced to below half of that of the UE. This confirms that simple low-temperature extractions cannot extract substances with the targeted activity; however, the UE extraction produces higher concentrations than the concentrations ranging from 8 to 15 mg/g reported in other works using chlorophyll extractions with other solvents, process temperatures, or complex processes in *Spirulina* [33,38]. Therefore, we could conclude that *Spirulina* extract from UE results in relatively high extract yields, and the activity of the targeted bioactive substances is maintained. These results were similar to other studies that mentioned the excellence of UE in low-temperature extraction processes [39,48].

Table 1. Estimation of the concentrations of total chlorophylls and extraction yields of the *Spirulina* extracts from two different extraction processes.

Extraction Process	Total Chlorophyll Contents (mg/g)	Extraction Yields (%)
Conventional ethanol extraction (EE) *	6.24 ± 0.92	8.9 ± 1.66
Ultrasonic extraction (UE) **	17.56 ± 1.86	11.6 ± 2.02

* EE, 70% ethanol extraction at 80 °C for 12 h. ** UE, ultrasonic pretreatment with 70% ethanol at 40 kHz and room temperature for 8 h, and further extraction at 65 °C for 4 h.

As shown in Table 1, the extracts from the UE are expected to contain intact forms of chlorophylls at high concentrations and have many beneficial biological activities. In particular, 70–80% of chlorophyll in *Spirulina* is known to be in a form of chlorophyll a, which reportedly has stronger antioxidant activity than other chlorophylls [33,34]. Therefore, the extract obtained through this process is expected to have strong antioxidative activity. The extract has also been reported in various to protect cranial nerves. However, most of these studies showed that the antioxidant activity amplifies the p-ERK/p-CREB/BDNF signaling pathway and enhances cognitive functions through inhibition of the acetylcholinesterase (AChE) enzyme [10,32,35]. Furthermore, studies on the anti-inflammatory action of the *Spirulina* extracts, or their chlorophylls on the resultant protection of nerve cells, are very rare. Therefore, the effects of the *Spirulina* extracts, obtained from the ultrasonic process on suppressing the inflammation of mouse nerve cells through antioxidation, are shown as follows.

3.2. Inhibition of Nitric Oxide (NO) and Prostaglandin E2 (PGE2) Production by the Extracts

Prior to carrying out the anti-inflammatory experiments, in Figure 1 the cell cytotoxicity of the samples against mouse microglial cells was compared by treating the UE or EE (0.01 to 0.1 mg/mL) with (black bars) or without 1 µg/mL of LPS (white bars).

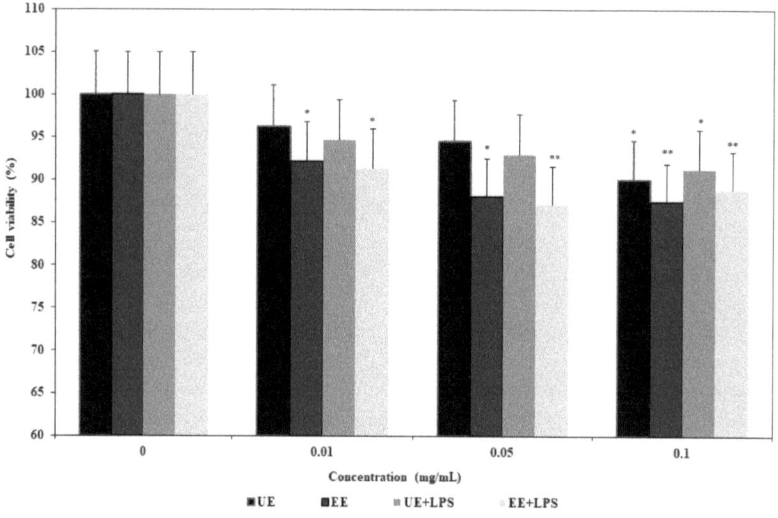

Figure 1. Cytotoxicity of the *S. maxima* extracts with (black bars) and without (white bars) adding 1 µg/mL of lipopolysaccharide (LPS) against BV-2 cells. EE, 70% ethanol extraction at 80 °C for 12 h, UE, ultrasonic pretreatment with 70 % ethanol at 40 kHz and room temperature for 8 h, and further extraction at 65 °C for 4 h. Values are presented as means ±SD; * $p < 0.05$ and ** $p < 0.01$ compared with the non-treatment group.

Interestingly, the EE from the hot 70% ethanol extraction process generally had higher cytotoxicity than the UE both with and without LPS, which was presumably attributable to the high heat-mediated degeneration of chlorophyll and resultant increases in toxic substances, such as pheophorbide. The effects of the samples on cell death also showed concentration dependency. The highest cell cytotoxicity (approximately 14%) was observed upon treatment of 0.1 mg/mL EE with LPS, compared to approximately 12% upon treatment of the UE at the same concentration; approximately 9% and 5% of the cells died upon treatment of 0.01 mg/mL UE and EE with LPS, respectively. In general, less cell cytotoxicity was observed in both samples without LPS, but the difference in the cytotoxicities of extracts treated with and without LPS was not significant. Therefore, the extracts from both processes could be used to investigate their anti-inflammatory effects because ca. 90% and 86% of the lowest survival rates were observed at the maximum dosage of 0.1 mg/mL without and with LPS, respectively. Figures 2 and 3 demonstrate the NO and PGE2 measurements after treatment with the two extracts in BV-2 cells. In Figure 2, the amount of NO produced in the untreated group not treated with LPS was extremely small at 5.21 µM, but the amount of NO produced in the LPS-treated group sharply increased to 40.31 µM. In contrast, when BV-2 cells were treated with 70% ethanol extract together with LPS, 35.6 µM NO was generated at the 0.01 mg/mL EE concentration, while 20.3 µM of NO was generated at the maximum concentration of 0.1 mg/mL. Additionally, when BV-2 cells were treated with UE extracts, 31.4 µM NO was generated at the low concentration, and 19.6 µM NO was generated at the high concentration, indicating that the extracts obtained from the UE had greater ability to suppress NO generation than extracts from conventional extraction processes. Additionally, both extracts suppressed the generation of NO in a concentration-dependent manner, demonstrating that UEs exert remarkable anti-inflammatory effects on brain nerve cells and that 70% EEs also exhibit anti-inflammatory activity in brain nerve cells. These results are consistent with the findings of other studies [49], indicating that chlorophyll has anti-inflammatory effects and strongly suggesting that the anti-inflammatory effect of *Spirulina* extract is attributable to chlorophyll; similar results regarding the antioxidant effect of chlorophyll were also reported in other studies [37,50]. Additionally, when the extracts were administered alone at high concentrations, NO was minimally produced to levels similar to the control group not treated with LPS (data not shown).

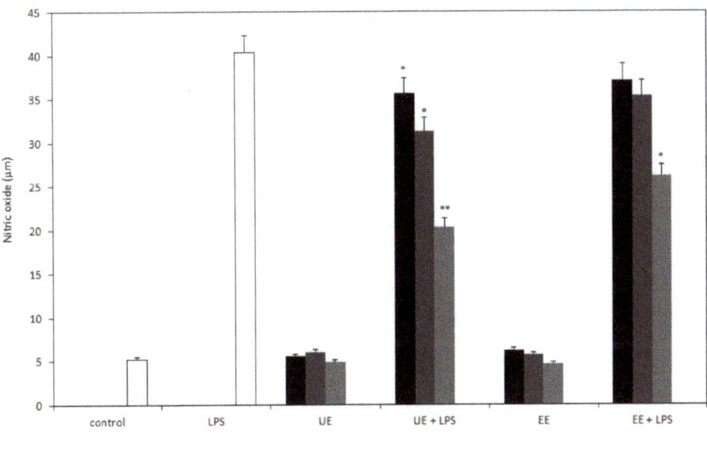

Figure 2. Secretion of nitric oxide from BV-2 cells by the treatment of various concentrations of the *S. maxima* extracts. EE, 70% ethanol extraction at 80 °C for 12 h, UE, ultrasonic pretreatment with 70% ethanol at 40 kHz and room temperature for 8 h, and further extraction at 65 °C for 4 h. Values are presented as means ±SD; * $p < 0.05$ and ** $p < 0.01$ compared with the LPS group.

Figure 3 demonstrates the production of PGE2, a major product of phlogogenic mechanisms.

As shown in Figure 3, 88.4 pg/mL PGE2 was produced in the control without any treatment, while 1580.9 pg/mL PGE2 was produced when BV2 cells were treated with LPS, indicating that inflammation was induced. When the extracts were administered at concentrations ranging from 0.01 to 0.1 mg/mL, PGE2 production decreased in a concentration-dependent manner in response to both extracts. Approximately 826.5 pg/mL PGE2 was produced when the UE was administered at a maximum concentration of 0.1 mg/mL. In general, the amount of PGE2 produced decreased according to concentration in the same pattern as that of NO production. The ultrasonic extract had a more dramatic reduction compared to the extract obtained by the general extraction process, reconfirming that ultrasound extracts have stronger anti-inflammatory effects. When compared to *Coridalyis bungeana* [51], *Coptids rhizome* [52], and Royal jelly [53], the *Spirulina* extracts from this UE showed anti-inflammatory effects selectively on brain nerve cells similar to or higher than those of the other substances. The extract containing large amounts of the useful bioactive substance in intact forms without the destruction of activity obtained through ultrasonic extraction has better anti-inflammatory properties, therefore confirming the excellence of ultrasonic low-temperature extraction once again. In particular, NO is known to be excessively produced by macrophages in response to stimulation with substances, such as LPS and amyloid-beta which are toxic intracellular substances to cause cytotoxicity and inflammation [54]. PEG2 is synthesized by COX-2 to mediate the pain and fever on damaged tissues or cell regions and is known to be involved in the induction of Parkinson's and Alzheimer's disease at high levels [55]. Therefore, effective inhibition of these substances may reduce factors that cause inflammation in brain nerve cells, thereby improving cognitive functions through anti-inflammatory effects. This information can be understood in the same context as previously reported findings, indicating that *Spirulina* can protect brain nerve cells and improve cognitive function and memory [31,32,35].

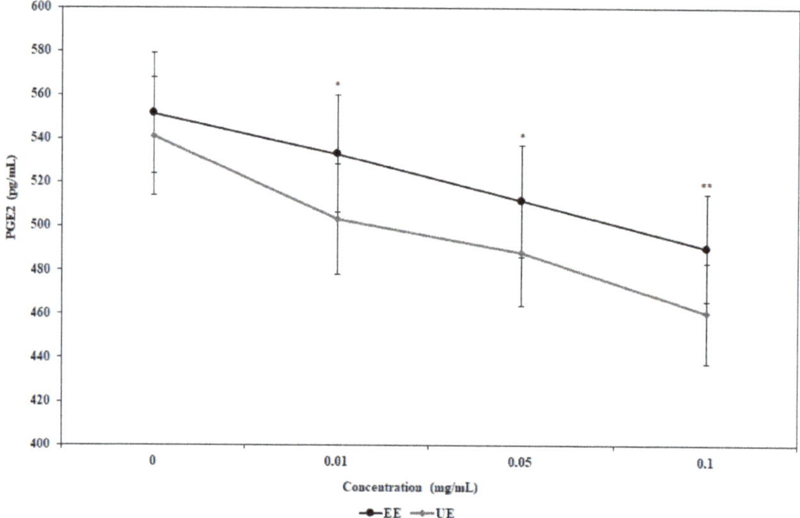

Figure 3. Comparison of PGE2 secretion from BV-2 cells by the treatment of various concentrations of the *S.maxima* extracts. EE, 70% ethanol extraction at 80 °C for 12 h, UE, ultrasonic pretreatment with 70% ethanol at 40 kHz and room temperature for 8 h, and further extraction at 65 °C for 4 h. Values are presented as means ±SD; * $p < 0.05$ and ** $p < 0.01$ compared with the non-treatment group.

3.3. Inhibition of mRNA Expression and Secretion of Pro-Inflammatory Cytokines by the Extracts

Because the *Spirulina* extracts proved to be capable of suppressing the production of mediators that directly affect inflammation, Figures 4–7 show the effects of the two extracts on the mRNA expression of TNF-α, IL-6 and IL-1β, which are pro-inflammatory cytokines that directly affect the inflammation of brain nerve cells. These experiments also assess the degree to which the extracts suppress the actual production of these cytokines in mouse nerve cells.

Figure 4 shows RT-PCR products that reflect the expression of mRNA in the cells treated with only LPS, cells treated with both LPS and extracts, and the control with no not treatment. As shown in Figure 4a, the amount of TNF-α increased sharply in the mouse microglial cells treated with LPS alone compared to the control without any treatment. In contrast, when LPS and *Spirulina* extracts were administered together, TNF-α expression decreased in a concentration-dependent manner. Similar to the effects on NO and PGE2 production, shown in Figures 2 and 3, the ultrasonic extract had greater inhibitory effects on expression than the extract obtained through the general extraction process. To quantitatively compare the electrophoretic bands, shown in Figure 4a, the sizes of individual bands were normalized to the beta-actin band, which is a house-keeping gene, using a program that quantitatively compares the band sizes, as shown in Figure 4b. Similar to the pattern in Figure 4a, mRNA expression relative to beta-actin decreased as the concentration of the extracts administered increased. Additionally, the ultrasonic extract had much greater inhibitory activity than the general extract, as shown in Figure 4a. In particular, the difference in TNF-α expression inhibition between the two extracts was larger higher extract concentrations of 0.1 mg/mL than at lower concentrations, possibly, due to higher amounts of chlorophylls in the extracts. Therefore, this result indicates once again that extraction of bioactive substances sensitive to heat requires a low-temperature extraction process and that the UE is the most efficient extraction process in such cases.

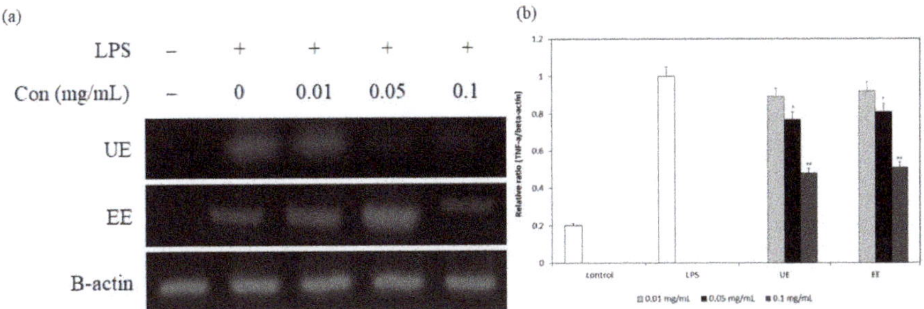

Figure 4. Down-regulation of mRNA expression of TNF-a from LPS-induced BV-2 cells (**a**) and the relative ratio of the gene expression by normalizing with beta-actin as a house-keeping gene (**b**) by the treatment of various concentrations of the *Spirulina* extracts along with the untreated control. EE, 70% ethanol extraction at 80 °C for 12 h, UE, ultrasonic pretreatment with 70% ethanol at 40 kHz and room temperature for 8 h, and further extraction at 65 °C for 4 h. Values are presented as means ±SD; $*$ $p <$ 0.05 and $**$ $p <$ 0.01 compared with the LPS group.

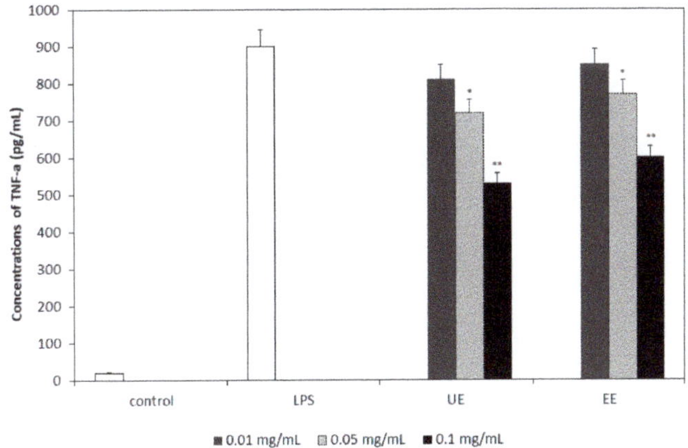

Figure 5. The secretion of TNF-a from LPS-induced BV-2 cells by the treatment of various concentrations of the *Spirulina* extracts. EE, 70% ethanol extraction at 80 °C for 12 h, UE, ultrasonic pretreatment with 70% ethanol at 40 kHz and room temperature for 8 h, and further extraction at 65 °C for 4 h. Values are presented as means ±SD; $*$ $p <$ 0.05 and $**$ $p <$ 0.01 compared with the LPS group.

As shown in Figure 5, the secretion of TNF-α from BV-2 cells was observed to confirm whether down-regulation of mRNA expression actually reflects the suppression of TNF-α secretion in mouse central nervous system (CNS) nerve cells. Similar to the pattern of the transcription of mRNA in Figure 4, the secretion of TNF-α was also decreased as a concentration-dependent manner with the greatest reductions at high concentrations. Unusually, although the number of general extracts produced concentration-dependent reductions, the reductions were relatively smaller compared to the ultrasonic extract. Although mRNA expression suppression, as shown in Figure 4, appeared to be concentration-dependent for both extractions, the degree of TNF-α suppression was lower in the general extract treatment. This means that mRNA expression is more sensitive to the useful components in the extract, but dramatic effects occur only when the concentration of useful substances is high enough in the following transcription stage. Therefore, we hypothesize that to anticipate intracellular inhibition or enhance functions using natural products along with gene expression, the natural products should exist at least at the critical concentration necessary to affect the production of the target substance. Figures 6 and 7 show the quantification of IL-6 and IL-1β gene expression, which was determined using the Image J program with the IL-6 and IL-1β RT-PCR product bands, as shown in Figure 4a, to more easily compare the suppression levels of these genes than that allowed by the pictures of electrophoresis bands themselves. Figure 6a shows the quantitative comparison of the RT-PCR-amplified bands reflecting mRNA levels of the target and housekeeping genes after treatment with LPS and the two extracts at different concentrations. Figure 7a shows the comparison of IL-1β mRNA gene expression levels.

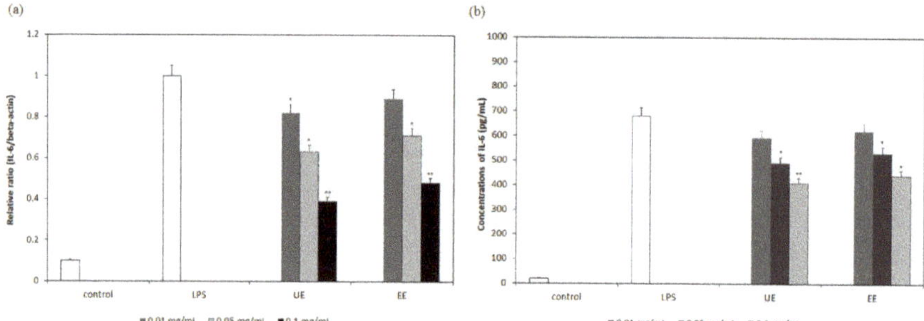

Figure 6. Relative ratios of mRNA expression of IL-6 by normalizing with beta-actin as a house-keeping gene (**a**) and secretion of IL-6 (**b**) from LPS-induced BV-2 cells by the treatment of various concentrations of the *Spirulina* extracts. EE, 70% ethanol extraction at 80 °C for 12 h, UE, ultrasonic pretreatment with 70% ethanol at 40 kHz and room temperature for 8 h, and further extraction at 65 °C for 4 h. Values are presented as means ±SD; * $p < 0.05$ and ** $p < 0.01$ compared with the LPS group.

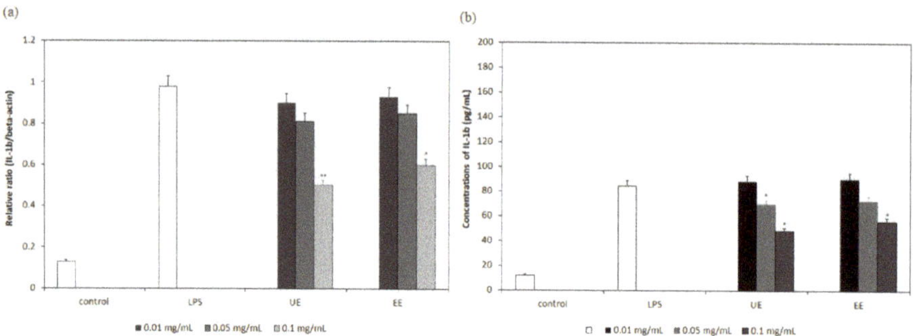

Figure 7. Relative ratios of mRNA expression of IL-1beta by normalizing with beta-actin as a house-keeping gene (**a**) and secretion of IL-6 (**b**) from LPS-induced BV-2 cells by the treatment of various concentrations of the *Spirulina* extracts. EE, 70% ethanol extraction at 80 °C for 12 h, UE, ultrasonic pretreatment with 70% ethanol at 40 kHz and room temperature for 8 h, and further extraction at 65 °C for 4 h. Values are presented as means ±SD; * $p < 0.05$ and ** $p < 0.01$ compared with the LPS group.

These two graphs do not show the electrophoretic bands, as shown in Figure 4a, to avoid redundancy, as the graphs in Figures 6a and 7a show the quantitative comparison of the bands obtained by electrophoresis. Examination of IL-6 and IL-1β gene expression after extract administration in Figures 6a and 7a revealed that the degree of IL-1β inhibition is generally higher than that of IL-6 for both extracts, although they are similar. This suggests that *Spirulina* extract may act more selectively on IL-1β than on IL-6 transcript levels. According to this difference of IL-6 and IL-1β protein secretion, shown in Figures 6b and 7b, the amount of IL-6 secreted, which was 680.5 pg/mL when 0.01 (mg/mL) of the ultrasonic extract was administered, decreased by 29.4% to 480.3 pg/mL when 0.1 mg/mL was administered. The amount of IL-1β secretion decreased by approximately 47% (from 94.5 pg/mL to 48.2 pg/mL), indicating that the extract from UE is more effective at inhibiting IL-1β and TNF-α production among the pro-inflammatory cytokines. Moreover, similar to the trend for the suppression of TNF-α production, the UE showed higher inhibitory effects than the EE from conventional hot ethanol extraction processes, demonstrating for the first time that ultrasonic extracts are generally more effective at inhibiting inflammation of mouse nerve cells than the extract from the hot extraction process. Although the inhibitory activity was generally lower than that of the ultrasonic extracts, the *Spirulina* extract from conventional 70% ethanol extraction also inhibited the secretion of pro-inflammatory cytokines from mouse nerve cells to levels similar to or higher than those from other natural products [5,9,15,16].

3.4. Correlation Between Gene Down-Regulation and Secretion of Pro-Inflammatory Cytokines

The above-mentioned results suggest that the extracts down-regulate the expression of the genes involved in pro-inflammatory cytokine production, thereby inhibiting the secretion of the relevant cytokines. Many other studies have reported similar results. However, there are no published data on the quantitative correlation between gene regulation and protein secretion, since it is considered to be natural that the amounts of the cytokines secreted from BV-2 cells measured by ELISA analysis only support the results of gene expression from PCR analysis.

Therefore, Figure 8 shows the comparison of the down-regulation of the expression of TNF-α (Figure 8a), IL-6 (Figure 8b), and Il-1β (Figure 8c) genes, which are the cytokines most closely involved in the inflammation of brain nerve cells and resultant secretion of cytokines according to two different extraction processes, such as ultrasonic and conventional extraction processes. As shown in Figure 8a, for TNF-α, gene expression and protein production decrease proportionally according to the concentrations of extracts administered with high correlations. In particular, the degree of downregulation of gene expression and TNF-α production is constant at all concentrations of the extracts without any large difference among concentrations. This means that the extracts act on gene regulation the most directly leads to the control of TNF-α production. Therefore, it can be inferred that *Spirulina* extracts inhibit TNF-α production directly from transcription to induce anti-inflammatory effects. In contrast, for IL-6 (Figure 8b), although gene expression and IL-6 production inhibition occurred, the decreasing rates were not significantly proportional to each other. In particular, the analysis indicated that the degree of IL-6 production inhibition was higher than the degree of gene expression inhibition indicating the possibility that inhibition of IL-6 production from BV-2 cells occurs through pathways than gene expression or compositely unlike the TNF-α. Unlike the previous examples, IL-1β shows higher correlations between gene expression and inhibition of IL-6 production, but it has lower correlations than TNF-α. Therefore, it can be assumed that the extracts first control the amplification of TNF-α and IL-1β genes thereby directly inhibiting the production of cytokines. Additionally, the production of IL-6 is more effectively reduced because of the effects on the IL-6-producing gene and the reduction of TNF-α and IL-1β production, which have already been inhibited. The basic data on these mutual relationships have been presented for the first time in this study, and more detailed studies on the mutual relationships between *Spirulina* extracts and the three pro-inflammatory cytokines are needed to confirm the findings. However, the data shows for the first time that although *Spirulina* extracts are involved in inhibiting the secretion of three cytokines, rather than acting simultaneously, there are differences in the degree of inhibition mechanisms, and order of action. Therefore, these results suggest that rather than uniformly influencing the secretion of all pro-inflammatory cytokines, the natural extracts selectively affect the production. Future studies on the anti-inflammatory effects of the extracts are necessary to investigate more subdivided inhibition of target cytokines. In particular, stimulation by LPS is most directly involved in inflammatory diseases, such as septic shock, rheumatoid arthritis, insulin resistance, and cachexia, by maximizing TNF-α production from macrophages [55,56]. Therefore, since the ultrasonic extract inhibits TNF-α secretion most effectively, the anti-inflammatory effects of these extracts are very high. In particular, the effects of the *Spirulina* extracts, containing chlorophyll on the inflammatory mediator and inflammatory cytokine production inhibition, were shown. This strongly implies that in addition to the antioxidant effects of chlorophyll and *Spirulina* extracts that are already known, the extracts also have an ability to protect against nerve cell inflammation. Therefore, the neuroprotective effects of *Spirulina* extracts were caused by inhibiting the inflammation of mouse nerve cells via antioxidant properties, potentially improving cognitive activities. The results would suggest another possible mechanism of the *Spirulina* extracts on its neuroprotection process.

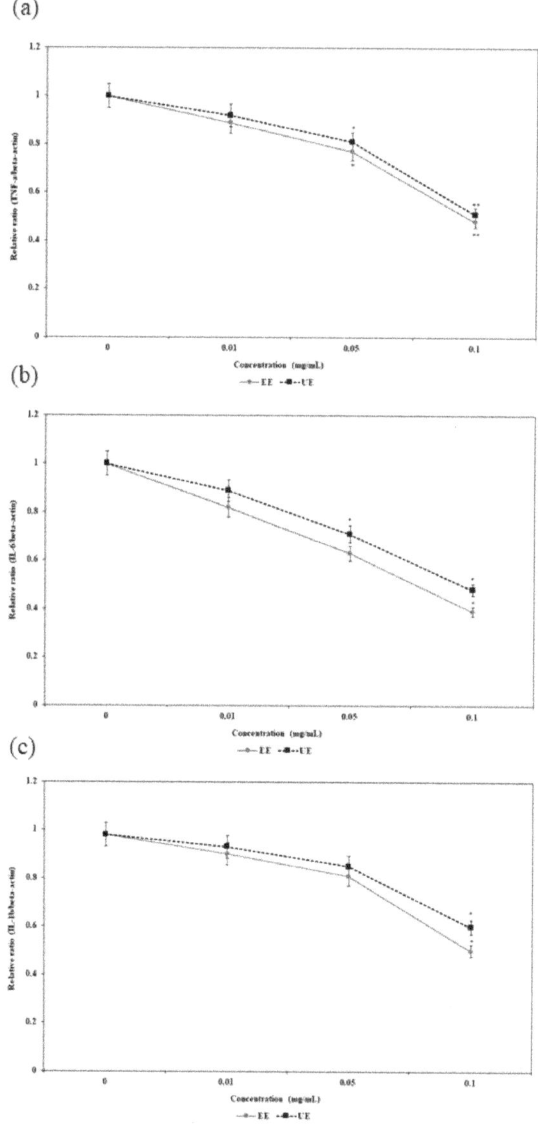

Figure 8. Quantitative comparison of the secretion and mRNA expression of TNF-a (**a**), IL-6 (**b**) and IL-1beta (**c**). EE, 70% ethanol extraction at 80 °C for 12 h, UE, ultrasonic pretreatment with 70% ethanol at 40 kHz and room temperature for 8 h, and further extraction at 65 °C for 4 h. Values are presented as means ±SD; * $p < 0.05$ and ** $p < 0.01$ compared with the non-treatment group.

4. Conclusions

This work first showed that the *Spirulina* extract from the UE enhanced anti-inflammation activities in mouse microglial cells associated with its antioxidant effects, compared with that from the conventional 70% ethanol extract at high temperature. The non-thermal ultrasonic extract contained more than two-fold higher amounts of extremely heat-sensitive chlorophylls than the conventional extract and exhibited a two-fold higher extraction yield. Specifically, the extract effectively inhibited the production of inflammation mediators, such as NO and PGE2, while the conventional extract also showed a relatively high ability to inhibit their production. Strong anti-inflammatory effects of the extract from the UE were also demonstrated by the significant down-regulation of mRNA expression of the pro-inflammatory cytokines, TNF-α, IL-6 and IL-1β. Moreover, we quantitatively demonstrated that inhibition of the gene expression was closely related to the suppression of pro-inflammatory cytokine secretion even though ELISA analysis supports PCR results in general; in particular, TNF-α had a very strong relationship while IL-6 demonstrated the least correlation between the down-regulation of gene expression and the suppression of cytokine secretion. Notably, the effect on TNF-α expression was greater than that on IL-1β and IL-6, which changed the least. This result suggests that the extract first controls TNF-α secretion and later inhibits IL-6, but this hypothesis should be further validated with a more detailed mechanism of the anti-inflammation cascade within nerve cells. However, we clearly showed that the extract from the ultrasonic process was more effective at suppressing the secretion of both inflammatory mediators and pro-inflammatory cytokines, as reflected by its neuroprotective activities. Conclusively, the high anti-inflammatory effects of the extracts were closely correlated with the high amounts of chlorophylls, which have strong antioxidant activities, obtained only through a non-thermal ultrasonic process. The high anti-inflammatory effects of the *Spirulina* extract with mouse microglia cell line could be employed to develop a possible mechanism for in vivo anti-inflammation effects along with the results from primary nerve cells. However, the expression of anti-inflammatory cytokines, such as IL-10 and TGF-β, should also be studied to further elucidate more exact anti-inflammation effects of this extract. This study also provides useful information for developing functional foods from heat-labile natural resources.

Author Contributions: W.Y.C., J.-H.S., J.-Y.L. and D.H.K. carried out all of the experiments and H.Y.L. designed the whole experiments and drafted the manuscript. All authors read and approved the final manuscript.

Funding: This work has been supported by a research grant from the Marine Biotechnology Program funded by the Ministry of Oceans and Fisheries (PM61400), and all of the authors deeply appreciate their financial support.

Conflicts of Interest: The authors declare no conflict of interest.

References

1. Bartus, R.T.; Dean, R.L.; Beer, D.; Lipa, A.S. The cholinergic hypothesis of geriatric memory dysfunction. *Science* **1982**, *217*, 408–417. [CrossRef]
2. Citron, M. Alzheimer's disease: Treatments in discovery and development. *Nat. Neurosci.* **2002**, *5*, 1055–1057. [CrossRef]
3. Squire, L.R. Memory systems of the brain: A brief history and current perspective. *Neurobiol. Learn. Mem.* **2004**, *82*, 71–77. [CrossRef] [PubMed]
4. Minghetti, L. Role of inflammation in neurodegenerative diseases. *Curr. Opin. Neurol.* **2005**, *18*, 315–321. [CrossRef]
5. Asai, M.; Iwat, N.; Yoshikawa, A.; Aizaki, Y.; Ishiura, S.; Saido, T.C.; Maruyana, K. Berberine alters the processing of Alzheimer's amyloid precursor protein to decrease Abeta secretion. *Bioichem. Biophys. Res. Commun.* **2007**, *352*, 498–502. [CrossRef] [PubMed]
6. Floyd, R.A. Antioxidants, oxidative stress and degenerative neurological disorders. *Proc. Soc. Exp. Biol. Med.* **2001**, *222*, 236–245. [CrossRef]
7. Halliwell, B.; Gutterridge, J.M. Role of free radicals and catalytic metal ions in human diseases: An overview. *Methods Enzymol.* **1990**, *186*, 1–85. [CrossRef] [PubMed]

8. Boille, S.; Vande Vele, C.; Cleveland, D.W. ALS: A disease of motor neurons and their nonneuronal neighbors. *Neuron* **2006**, *52*, 130–144. [CrossRef]

9. Talhouk, R.S.; Karam, C.; Fostok, S.; El-Jouni, W.; Barbour, E.K. Anti-inflammatory bioactivities in plant extracts. *J. Med. Food* **2007**, *10*, 1–10. [CrossRef] [PubMed]

10. Jenkins, S.M.; Barone, S. The neurotoxicant trimethyltin induces apoptosis via caspase activation, p38 protein kinase and oxidative stress in PC12 cells. *Toxicol. Lett.* **2004**, *147*, 63–72. [CrossRef] [PubMed]

11. Ali, S.F.; LeBel, C.P.; Bondy, S.C. Reactive oxygen species formation as a biomarker of methylmercury and trimethyltin neurotoxicity. *Neurotoxicology* **1992**, *13*, 637648. [PubMed]

12. Corvino, V.; Marches, E.; Michetti, F.; Geloso, M.C. Neuroprotective strategies in hippocampal neurodegeneration induced by the neurotoxicant trimethyltin. *Neurochem. Res.* **2013**, *38*, 240–253. [CrossRef] [PubMed]

13. Chen, A.; Xiong, L.J.; Tong, Y.; Mao, M. The neuroprotective roles of BDNF in hypoxic ischemic brain injury. *Biomed. Rep.* **2013**, *1*, 167–176. [CrossRef] [PubMed]

14. Dawson, G.R.; Iversen, S.D. The effects of novel cholinesterase inhibitors and selective muscarinic receptor agonists in tests of reference and working memory. *Behav. Brain Res.* **1993**, *57*, 143–153. [CrossRef]

15. Zhang, W.; Zhang, X.; Zou, K.; Wang, J.L.; Wang, Y. Seabackthorn berry polysaccharide protects against carbon tetrachloride-induced hepatoxicity in mice via anti-oxidative and anti-inflammatory activities. *Food Funct.* **2017**, *8*, 3130–3138. [CrossRef]

16. Yuk, H.J.; Lee, J.W.; Park, H.A.; Seo, K.S.; Oh, S.R.; Ryu, H.W. Protective effects of coumestrol on lipopolysaccharide-induced acute lung injury via the inhibition of pro-inflammatory mediators and NF-kB activation. *J. Funct. Food* **2017**, *34*, 181–188. [CrossRef]

17. Young, I.S.; Woodside, J.V. Antioxidants in health and disease. *J. Clin. Pathol.* **2001**, *54*, 176–186. [CrossRef]

18. Bennett, P.B.; Marquis, R.E.; Demchenko, I. *High Pressure Biology and Medicine*; Bennett, P.B., Ed.; University of Rochester Press: New York, NY, USA, 1998; pp. 1–23.

19. Lim, G.P.; Chu, T.; Yang, F.; Beech, W.; Frautschy, S.A.; Cole, G.M. The curry spice curcumin reduces oxidative damage and amyloid pathology in an Alzheimer transgenic mouse. *J. Neurosci.* **2001**, *21*, 8370–8377. [CrossRef]

20. Murabe, Y.; Sano, Y.; Roisen, F.J. Morphological studies on neuralgia. VI. Postnatal development of microglial cell. *Tissue Res.* **1982**, *225*, 469–485. [CrossRef]

21. Nakajima, K.; Kohsaka, D.; Raquel, E.G. Functional roles of microglia in the brain. *Neurosci. Res.* **1993**, *17*, 187–203. [CrossRef]

22. Gonzalez-Scarano, F.; Baltuch, G. Microglia as mediators of inflammatory and degenerative diseases. *Annl. Rev. Neurosci.* **1999**, *22*, 219–240. [CrossRef]

23. Witkamp, R.; Monshouwer, M. Signal transduction in inflammatory processes, current and future therapeutic targets: A mini review. *Vet. Q.* **2002**, *22*, 11–16. [CrossRef] [PubMed]

24. Eikelenboom, P.; van Gool, W.A. Neuroinflammatory perspectives on the two faces of Alzheimer's disease. *J. Neural Transm.* **2004**, *11*, 281–294. [CrossRef] [PubMed]

25. Liu, J.S.; Hong, J.S. Role of microglia in inflammation-mediated neurodegenerative diseases: Mechanisms and strategies for therapeutic intervention. *J. Pharmacol. Exp. Ther.* **2003**, *304*, 1–7. [CrossRef]

26. Gao, H.M.; Liu, B.; Zhang, W.; Hong, J.S. Novel anti-inflammatory therapy for Parkinson's disease. *Trends Pharmacol. Sci.* **2003**, *24*, 395–401. [CrossRef]

27. Saeid, A.; Chojnacka, K.; Korczyński, M.; Korniewicz, D.; Dobrzański, Z. Biomass of *Spirulina maxima* enriched by biosorption process as a new feed supplement for swine. *J. Appl. Phycol.* **2013**, *25*, 667–675. [CrossRef]

28. Son, M.H.; Park, K.H.; Choi, A.R.; Yoo, G.; In, M.J.; Kim, D.H.; Chae, H.J. Investigation of biological activities of enzymatic hydrolysate of *Spirulina*. *J. Korean Soc. Food Sci. Nutr.* **2009**, *38*, 136–141. [CrossRef]

29. Hernandez, A.C.; Nieves, I.; Meckes, M.; Chamorro, G.; Barron, B.L. Antiviral activity of *Spirulina maxima* against Herpes simplex virus type 2. *Antivir. Res.* **2002**, *56*, 279–285. [PubMed]

30. Zao, X.; Yang, Y.H.; Cho, Y.S.; Chun, H.K.; Song, K.B.; Kim, M.R. Quality characteristics of *Spirulina maxima* added salad dressing. *East Asian Soc. Diet. Life* **2005**, *15*, 292–299.

31. Deng, R.; Chow, T.J. Hypolipidemic, antioxidant, and anti-inflammatory activities of microalgae *Spirulina*. *Cardiovasc. Ther.* **2010**, *28*, 33–45. [CrossRef]

32. Koh, E.J.; Seo, Y.J.; Choi, J.; Lee, H.Y.; Kang, D.H.; Kim, K.J.; Lee, B.Y. *Spirulina maxima* extract prevents neurotoxicity via promoting activation of BDNF/CREB signaling pathways in neuronal cells and mice. *Molecules* **2017**, *22*, 1363. [CrossRef] [PubMed]

33. Bhat, S. R Chlorophyll: The wonder pigment. *Sci Rep.* **2005**, *42*, 29–32.

34. Hosikian, A.; Lim, S.; Halim, R.; Danquah, M.K. Chlorophyll Extraction from Microalgae: A Review on the Process Engineering Aspects. *Int. J. Chem. Eng.* **2010**, *29*, 1–11. [CrossRef]

35. Koh, E.J. The Improvement of Cognitive Impairment and Neuroprotection Effects of *Spirulina maxima* Extract. Ph.D. Thesis, Cha University, Pangyo, Korea, 2018; pp. 1–125.

36. Kay, R.A.; Barton, L.L. Microalgae as food and supplement. *Crit. Rev. Food Sci. Nutr.* **1991**, *30*, 555–573. [CrossRef] [PubMed]

37. McCarty, M.F. The chlorophyll metabolite phytanic acid is a natural rexinoid—Potential for treatment and prevention of diabetes. *Med. Hypotheses* **2001**, *56*, 217–219. [CrossRef]

38. Lee, Y.J.; Kim, S.H.; Kim, J.S.; Han, J.A.; Seo, H.J.; Lim, H.J.; Choi, S.Y. Studies on Simultaneous Determination of Chlorophyll a and b, Pheophorbide a, and β-Carotene in *Chlorella* and *Spirulina* Products. *J. Food Hyg. Saf.* **2005**, *20*, 141–146.

39. Kim, W.I.; Chung, K.W.; Hong, I.K.; Park, K.A. Ultrasound Energy Effect on Solvent Extraction of Amaranth Seed Oil. *J. Korean Ind. Eng. Chem.* **2001**, *12*, 307–311, OSTI ID:20268268.

40. Oh, S.H.; Ahn, J.H.; Kang, D.H.; Lee, H.Y. The effect of ultrasonificated extracts of *Spirulina maxima* on the anticancer activity. *Mar. Biotechnol.* **2011**, *13*, 205–214. [CrossRef]

41. Lee, D.H.; Hong, J.H. Antioxidant activities of chlorella extracts and physicochemical characteristics of spray-dried *Chlorella* powders. *Korean J. Food Preserv.* **2015**, *22*, 591–597. [CrossRef]

42. Mosmann, T. Rapid colorimetric assay for cellular growth and survival: Application to proliferation and cytotoxicity assays. *J. Immunol. Met.* **1983**, *65*, 55–63. [CrossRef]

43. Lee, Y.H.; Jeon, S.H.; Kim, S.H.; Kim, C.Y.; Lee, S.J.; Koh, D.S.; Lim, Y.H.; Ha, K.S.; Shin, S.Y. A new synthetic chalcone derivative, 2-hydroxy-3′,5,5′-trimethoxychalcone (DK-139), suppresses the Toll-like receptor 4-mediated inflammatory response through inhibition of the Akt/NF-κB pathway in BV-2 microglial cells. *Exp. Mol. Med.* **2012**, *44*, 369–377. [CrossRef]

44. Kim, J.S.; Seo, Y.C.; Choi, W.Y.; Kim, H.S.; Kim, B.H.; Shin, D.H.; Lee, H.Y. Enhancement of antioxidant activities and whitening effect of Acer mono Sap through nano-encapsulation processes. *Korean J. Med. Corp. Sci.* **2011**, *19*, 191–197. [CrossRef]

45. Green, L.C.; Wagner, D.A.; Glogowski, J.; Skipper, P.L.; Wishnok, J.S.; Tannenbaum, S.R. Analysis of nitrate, nitrite, and [15N] nitrate in biological fluids. *Anal. Biochem.* **1982**, *126*, 131–138. [CrossRef] [PubMed]

46. Deck, L.M.; Hunsaker, L.A.; Gonzales, A.M.; Orlando, R.A.; Vander, D.L. Substituted trans-stilbenes can inhibit or enhance the TPA-induced up-regulation of activator protein-1. *BMC Pharmacol.* **2008**, *8*, 19–25. [CrossRef]

47. Kim, M.R.; Kang, O.H.; Kim, S.B.; Kang, H.J.; Kim, J.E.; Hwang, H.C.; Kim, I.W.; Kwon, D.Y. The study of anti-inflammatory effect of *Hwanggeumjakyak-tang* extract in RAW 264.7 Macrophage. *Korean J. Herbol.* **2013**, *28*, 43–50. [CrossRef]

48. Zheng, H.; Yin, J.; Gao, Z.; Huang, H.; Ji, X.; Dou, C. Disruption of *Chlorella vulgaris* Cells for the release of biodiesel-producing lipids: A Comparison of grinding, ultrasonication, bead milling, enzymatic lysis, and microwaves. *Appl. Biochem. Biotechnol.* **2011**, *164*, 1215–1224. [CrossRef]

49. Okamoto, K.; Kihira, T.; Kobashi, G.; Sasaki, S.; Yokoyama, T.; Inaba, Y.; Nagai, M. Fruit and vegetable intake and risk of amyotrophic lateral sclerosis in Japan. *Neuroepidemiology* **2009**, *3*, 251–258. [CrossRef]

50. Gruskin, B. Chlorophyll—Its therapeutic place in acute and suppurative disease: Preliminary report of clinical use and rationale. *Am. J. Surg.* **1940**, *49*, 49–55. [CrossRef]

51. Ju, M.K. Protorpine, and Alkaloid Isolated from *Croydalis bugeana* Turca, Attenuate Neuro-Inflammation in LPS-Induced BV-2 Microglia Cells. Master's Thesis, Kyungbuk National University, Daegu, Korea, 2017; pp. 1–56.

52. Akhter, M.H.; Sabir, M.; Bhide, N.K. Anti-inflammatory effect of berberine in rats injected locally with cholera toxin. *Indian J. Med. Res.* **1977**, *65*, 133–141. [PubMed]

53. Yu, M.M.; Chen, Y.F.; Pan, Y.M.; Liu, Y.C.; Tu, J.; Wang, K.; Hu, F.L. Royal jelly attenuate LPS-induced inflammation in BV-2 microglia cells through modulating NF-kB and p38/JNK signaling pathways. *Med. Inflamm.* **2018**, *2018*, 7834381. [CrossRef]

54. Abramson, S.B.; Attur, M.; Amin, A.R.; Clancy, R. Nitric oxide and inflammatory mediators in the perpetuation of osteoarthritis. *Curr. Rheumatol. Rep.* **2001**, *3*, 535–564. [CrossRef] [PubMed]
55. Kuo, C.L.; Chi, C.W.; Liu, T.Y. The anti-inflammatory potential of berberine in vitro and in vivo. *Cancer Lett.* **2004**, *203*, 127–137. [CrossRef] [PubMed]
56. Jhun, B.S.; Jin, Q.; Kin, Y.T.; Kong, S.S.; Cho, Y.; Ha, Y.H.; Baik, H.H.; Kang, I. 5-aminoinidazole-4-carboximide riboside suppresses LPS-induced TNF-alpha production through inhibition of posphophatidylinositol-3-kinase/A activation in RAW264.7 murine macrophage. *Biochem. Biophys. Res. Commun.* **2004**, *318*, 372–380. [CrossRef] [PubMed]

 *applied sciences*

Article

Psyllium and *Laminaria* Partnership—An Overview of Possible Food Gel Applications

Patrícia Fradinho [1,2,*], Anabela Raymundo [1], Isabel Sousa [1], Herminia Domínguez [2] and María Dolores Torres [2]

1. LEAF—Linking Landscape, Environment, Agriculture and Food, Instituto Superior de Agronomia, Universidade de Lisboa, 1349-017 Lisbon, Portugal; anabraymundo@isa.ulisboa.pt (A.R.); isabelsousa@isa.ulisboa.pt (I.S.)
2. Department of Chemical Engineering, Universidade de Vigo (Campus Ourense), Science Faculty, As Lagoas, 32004 Ourense, Spain; herminia@uvigo.es (H.D.); matorres@uvigo.es (M.D.T.)
* Correspondence: pfradinho@isa.ulisboa.pt; Tel.: +351-213653246

Received: 11 September 2019; Accepted: 13 October 2019; Published: 16 October 2019

Featured Application: *Laminaria-psyllium* **gels with distinct texture and rheological features, designed for a wide range of food applications.**

Abstract: Seaweeds are a novel source of important nutritional compounds with interesting biological activities that could be processed into added-value products. In this study, two previously developed products obtained by *Laminaria ochroleuca* processing (liquid extract and a purée-like mixture) were processed with *Psyllium* gel to develop functional hydrogels. The optimization of the formulation and the characterization of the *Laminaria-Psyllium* gels in terms of their mechanical features have allowed the proposal of potential food applications. A beneficial interaction was found between *Laminaria* and *Psyllium* in terms of the reinforcement of texture and rheological properties. The obtained outcomes could provide new healthy gelling formulations with attractive properties to alleviate the growing market demand of eco-novel food matrices.

Keywords: kombu; edible brown seaweed; gels; *Psyllium*; *Laminaria ochroleuca*; autohydrolysis; mechanical properties

---

1. Introduction

Laminaria sp. are industrially used for alginate extraction (17.1–32% *w/w*, dry basis) [1], a hydrocolloid with unique gelling abilities at low temperature and good heat stability, widely used as a thickener, stabilizer, and restructuring agent in the food, cosmetic, pharmaceutical, biomedical, and textile industries [2,3].

Mainly valued for alginate extraction, this natural resource presents other interesting compounds for human nutrition such as fatty acids, proteins, minerals [4], vitamins (A, C, D, B group, E, K, PP) [1], pigments such as carotenoids and polyphenols with proved antioxidant, hypoglycemic, antitumoral, and antimicrobial activities [5,6]. Due to their high content in biologically active compounds, seaweeds, and especially *Phaeophyceae* (brown algae), have great potential to act as a functional food and as a food ingredient [7,8].

Psyllium (*Plantago ovata*) is an annual plant found mainly in India and Pakistan, but also on Madeira Island (Portugal) due to its subtropical climate. Used in traditional medicine for centuries, *Psyllium* husk has gained more attention by the scientific community since its health allegation was approved by the Food and Drug Administration in 2012 [9] regarding its benefits in reducing the risk of coronary heart disease. These features come from its high soluble fiber composition, which has the capacity of absorbing up to 15 g of water per g of *Psyllium* [10].

Psyllium husk is composed of 85% arabinoxylan, a neutral highly branched polysaccharide with about 35% of non-reducing terminal residues and a 15% non-polysaccharide fraction [11]. It has applications as an edible coating [12] and drug delivery [13].

Innovation in food and feed production is a reality and a tendency for the upcoming years, whether this is the use of poorly exploited raw materials, the reformulation of foods based on green and sustainable technologies, or "back to tradition". Many of the food and feed industries' products are gels (e.g., yoghurt, puddings, confectionery products, pasta, pet food, among others) or present gelling agents in their formulation (cream cheese, sauces, …), making gel-systems a growing market.

Among the edible biopolymers, the use of plant seed mucilages namely chia and flax seed [14,15] are one of today's trends. Aside from the sustainability issue, these ingredients also have health benefits such as the regulation of colonic microbiota, reduction of hyperlipidemia, anti-inflammatory effect, control of glycemic response, and control of satiation [16]. Due to their technological properties, these biopolymers are often used in the food industry as texturizing agents.

To our knowledge, the gel forming ability of the combination of *L. ochroleuca-Psyllium* has not been previously investigated. Therefore, the interaction between *Psyllium* husk gel and the edible brown seaweed *Laminaria ochroleuca*, either in its liquid fraction or its purée-like mixtures, with *Psyllium* will be studied, taking advantage of both these poorly exploited natural resources. The aim of the present work is to lay the foundation for a systematic textural and rheological description of *Laminaria-Psyllium* gels, focusing on their mechanical features intended for future food applications.

2. Materials and Methods

2.1. Raw Materials

Dehydrated *Laminaria ochroleuca* (Algas Atlánticas Algamar, S.L., Pontevedra, Spain) was milled and sieved to two different particle size powders (0.25–2 mm and <0.25 mm diameters). *Psyllium* husk of Indian origin (Solgar, lot 107028-01, Leonia, NJ, USA) was purchased at the local market, milled (Pulverisette 14 Premium, Fritsch, Idar-Oberstein, Germany) and sieved to 160–315 μm.

For comparison purposes, commercial references were used, namely baby food (purée), guacamole (purée), fruit jam (gel), jelly gum (rubbery solid), pâté (liver paste), and pet food (moistened feed, pâté like).

2.2. Formulations Development

2.2.1. *Laminaria ochroleuca* Sample Production

Focusing on the full valorization of *L. ochroleuca*, the 2 mm fraction of the alga was subjected to autohydrolysis (AH) in a pressurized reactor (Parr Instruments series 4848, Moline, IL, USA) [14] and the liquid extract (liquor) obtained was stored at −18 °C until further use. When using the <0.25 mm alga fraction, purée-like mixtures were prepared following the procedure already described by the authors [17], with (LoUS) and without ultrasonic treatment (Lo). The alga purées were matured for 24 h at 4 °C before preparing the *Laminaria-Psyllium* gels.

2.2.2. Preparation of the *Laminaria-Psyllium* Gels

The control sample (Psy) was prepared by adding 4% (*w/w*, d.b) *Psyllium* husk (160–315 μm range of particle size) to distilled water at 40 °C under mechanical stirring (10 min, 200 rpm) [18]. Another *Psyllium* gel was prepared in a similar manner as the control by using the *Laminaria* autohydrolysis liquid extract (PsyL).

Laminaria-Psyllium gels were prepared by mixing (Eurostar Power-b, Ika-Werke, Staufen im Breisgau, Germany) for 5 min/200 rpm at room temperature, the control sample and the alga purée mixture at the ratios 25:75, 50:50, and 75:25 in order to obtain visually different structures with potential different applications. Two batches of 160 g of each gel (*Laminaria-Psyllium*: 25:75, 50:50, 75:25) were prepared, poured into sealed plastic containers (Ø = 10 cm) and stored at 4 °C for 24 h to maturate. The schematic procedure of the sample preparation and analysis is presented in Figure 1.

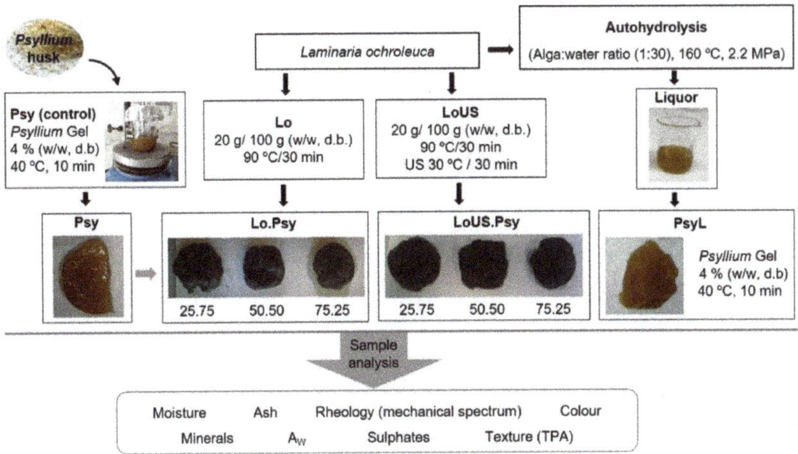

Figure 1. General schematic procedure of the gel preparation and analysis (Lo—*Laminaria ochroleuca*; Psy—*Psyllium* gel prepared with water (control); US—ultrasound treatment; PsyL—*Psyllium* gel prepared with autohydrolysis liquor).

2.3. Physicochemical Measurements

Moisture and ash contents of the raw materials and gels were determined by gravimetric methods by placing the samples in an oven (105 ± 2 °C) until constant weight, or in the furnace (575 °C, 6 h) respectively.

Sulfate content of the developed gel samples and *Psyllium* husk was obtained by the ionic chromatography method (mobile phase: 3.2 mM sodium carbonate/1 mM sodium bicarbonate at 0.70 mL/min) as previously reported [19].

Water activity (a_w) measurements of all gel samples were performed using a LabMaster-aw (Novasina, Pfäffikon, Switzerland) in triplicate.

Mineral content (Ca, K, Na, Fe) of *Psyllium* husk was analyzed by inductively coupled plasma optical emission spectrometry (Optima 4300 DV, Perkin Elmer, Waltham, MA, USA) after microwave-assisted (Savillex, Eden Prairie, MN, USA) acid digestion (80 °C, 6 h).

2.4. Color Measurements

The color evaluation was performed using a CR-400 colorimeter (Minolta, Tokyo, Japan) with standard illuminant C. Tristimulus color coordinates (CIELAB system) were used to measure the degree of lightness (L*), which ranged from 0 (black) to 100 (white), redness (a*) ranging from −60 (green) to +60 (red), and yellowness (b*) ranging from blue (−60) to yellow (+60). Total color difference (ΔE*) between the samples and the control was calculated according to Equation (1).

$$\Delta E^* = (\Delta L^* + \Delta a^* + \Delta b^*)^{1/2} \tag{1}$$

For calibration purposes, a white standard was used (L* = 97.21; a* = 0.14; b* = 1.99). Measurements were conducted at 20 ± 1 °C and replicated at least five times.

2.5. Dynamic Viscoelasticity

Dynamic oscillatory rheology measurements were used to monitor the viscoelastic characteristics of the gels. Small amplitude oscillatory shear measurements were conducted in triplicate in a controlled stress rheometer (RheoStress 600, Haake, Vreden, Germany) using serrated parallel plate geometry (diameter 35 mm) and a 1.5 mm gap. Surface geometry was covered with paraffin oil to prevent moisture loss. Samples were rested in the rheometer device for 5 min (20 °C) before rheological testing to temperature equilibration. Initially, stress sweep tests were run at 1 Hz, with the shear stress of the input signal varying from 0.1 to 100 Pa to find the linear viscoelastic region. The mechanical spectra of all samples were assessed through frequency sweep tests performed from 0.1 to 10 Hz (20 °C, 10 Pa) within the linear viscoelastic region previously defined. Experimental storage (G′) and loss (G″) moduli (Pa) data versus frequency (f, Hz) were fitted using well-known power models reported elsewhere [20], where α′, α″, b′, and b″ are the corresponding fitting parameters (Equations (2) and (3)).

$$G' \, (f) = \alpha' \, f^{b'} \tag{2}$$

$$G'' \, (f) = \alpha'' \, f^{b''} \tag{3}$$

2.6. Texture Profile Analysis (TPA)

Texture profile analysis of all developed gels was performed in a TA-XT2 texturometer (Stable Microsystems, Godalming, UK) with a P0.25S plunger that penetrated 15 mm into the sample at 1 mm/s. From the force vs. time texturogram, three parameters were selected to characterize the materials: firmness, as the maximum rupture force (N); adhesiveness, represented by the negative area of the graph, that translates the recessive force of the probe (N·s); and cohesiveness corresponding to the ratio of the material's response to a second deformation. At least five measurements were made in each sample. Commercial food and feed products were also measured for comparative purposes.

2.7. Statistical Analysis

Statistical analysis of the experimental data was performed using RStudio (Version 1.2.1335—© 2009–2019 RStudio, Inc., Boston, MA, USA), using variance analysis (one-way ANOVA), and the Tukey test, *Post Hoc* comparison at a significance level of 95% ($p < 0.05$). A Pearson correlation analysis was also conducted ($p < 0.05$) to determine the relationships between the color, texture parameters, and the amount of alga of the samples. The curve fitting of the rheological data was performed in Excel (version 365, Microsoft). All results are presented as the mean ± standard deviation (s.d.).

3. Results and Discussion

3.1. Physicochemical Characterization of Samples

Psyllium gel prepared with autohydrolysis (AH) liquor (PsyL) was visually more fluid than the *Psyllium* gel prepared with water (control). This is confirmed by texture and rheology measurements, so this issue will be discussed in Sections 3.3 and 3.4.

As seen in Figure 1, the *Laminaria-Psyllium* 75.25 gel samples presented syneresis (water in the filter paper, not measured) more evident in the sample not submitted to ultrasonic treatment. According to the previous study in which the purée-like mixtures were optimized [17], the authors did not find syneresis in the systems. This could be due to the alga high sodium content that bonds to *Psyllium*, and at the 25% ratio was not able to retain all the water in the system, causing the *Laminaria-Psyllium* 75.25 gels to contract and release part of the water previously enclosed. This phenomenon did not occur in samples with higher *Psyllium* gel ratios. A recent study by Figueroa and co-workers [21] reported the positive influence of *Psyllium* on the absence of syneresis of fruit jellies enriched with dietary fiber.

In Table 1, a chemical characterization of *Psyllium* husk and the developed gels is presented.

Table 1. Moisture, ash, and sulfate content of *Laminaria ochroleuca*, *Psyllium* husk, and the developed gels.

Samples	Moisture	Ash	Sulphates
	(%)	(%, d.b.)	
Laminaria ochroleuca [17]	9.20 ± 0.07	35.01 ± 0.31	2.21 ± 0.10
Psyllium husk (160–315 μm)	9.03 ± 0.31 h	2.98 ± 0.05 g	0.09 ± 0.00 e
Psy (control)	95.83 ± 0.08 a	2.91 ± 0.18 g	0.20 ± 0.00 e
PsyL	93.58 ± 0.06 b	18.01 ± 1.83 f	0.98 ± 0.00 d
Lo.Psy_25.75	91.73 ± 0.03 c	24.75 ± 0.51 d	1.75 ± 0.01 c
Lo.Psy_50.50	87.50 ± 0.06 e	33.08 ± 0.24 b	2.20 ± 0.07 a,b
Lo.Psy_75.25	84.48 ± 0.22 g	36.02 ± 0.23 a	1.98 ± 0.13 b,c
LoUS.Psy_25.75	91.66 ± 0.12 c	21.91 ± 0.71 e	1.78 ± 0.02 c
LoUS.Psy_50.50	87.78 ± 0.13 d	30.34 ± 0.41 c	2.14 ± 0.11 a,b
LoUS.Psy_75.25	85.18 ± 0.12 f	34.71 ± 0.37 a,b	2.25 ± 0.01 a

Psy (control), *Psyllium* gels prepared in water; PsyL, *Psyllium* gels prepared in autohydrolysis liquor; *Laminaria*-*Psyllium* gels: Lo.Psy_25.75, Lo.Psy_50.50, and LoPsy_75.25; *Laminaria* with ultrasonic treatment-*Psyllium* gels: LoUS.Psy_25.75, LoUS.Psy_50.50, and LoUSPsy_75.25. Data are presented as the mean ± sd. Different letters in a column show significantly different data values at the $p < 0.05$ level.

The moisture content of the *Laminaria*-*Psyllium* gels depends greatly on the proportion of both components, and decreased with the increase in *Laminaria* content. Ash content revealed the reverse trend, with the *Laminaria* proportion being crucial for the final gel mineral content. This is due to the high ash content (35%, d.b.) of this alga, as we previously determined [17]. It is also noteworthy that the ultrasonic treatment applied to the purée-like mixtures had a significant ($p < 0.05$) negative influence on the total ash content of the gel, especially in the 25.75 and 50.50 samples.

Sulfate presence in the *Laminaria*-*Psyllium* gels is a clear indication of the presence of fucoidan, a sulfated polysaccharide with reported activity against stomach-gastric adenocarcinoma cells and lung carcinoma cells [22].

The most important chemical feature of *Psyllium* husk is its high soluble fiber content (80%, d.b., as previously determined by the authors in the same sample [10]. However, aside from its fiber content, *Psyllium* husk has a mineral composition of 10.3 g K/kg, 0.88 g Na/kg, 1.36 g Ca/kg, and 94.7 mg Fe/kg, (present study), the last two minerals having a higher content than most cooked pulses [23].

Another important feature is *Psyllium*'s ability to retain sodium at physiological important conditions (pH 1.2—stomach; pH 6.8—intestine), being potentially active in reducing the bioavailable fraction of ingested sodium in the body [24].

Considering both the *L. ochroleuca* [17] and *Psyllium* husk's mineral content, one can conclude that the partnership between this alga and *Psyllium* husk could be advantageous for the development of new food products.

Water activity was very high and ranged between 0.999–1.000 for all samples (data not shown) and these are typical values for gels, although some reduction was expected in gels with a high proportion of *Psyllium*.

3.2. Color Evaluation of Samples

From the color evaluation results presented in Table 2, the *Psyllium* gels (control and PsyL) only differed in terms of their chromatic parameters, which could be of importance depending on the desired application.

Table 2. Color parameters (L*, a*, b*) and ΔE* of *Psyllium* husk, *Laminaria ochroleuca* and its AH liquor, and the gels developed.

Samples	L *	a *	b *	ΔE *
Psyllium husk (160–315 μm)	59.39 ± 0.91 a	6.06 ± 0.18 a	24.22 ± 0.37 a	-
Laminaria ochroleuca [17]	56.36 ± 0.57	−2.48 ± 0.04	14.49 ± 0.18	-
AH liquor [17]	31.42 ± 1.74	1.45 ± 0.26	2.88 ± 1.60	-
Psy (control)	33.12 ± 1.78 b,c	1.32 ± 0.14 c	9.85 ± 1.36 c	-
PsyL	35.70 ± 2.86 b	3.78 ± 1.50 b	17.67 ± 5.54 b	8.6
Lo.Psy_25.75	30.08 ± 0.37 c,d	−0.16 ± 0.22 d	12.82 ± 0.57 c	4.5
Lo.Psy_50.50	27.50 ± 0.77 d,e	−0.07 ± 0.16 d	12.37 ± 1.07 c	6.3
Lo.Psy_75.25	25.14 ± 1.70 e	−0.45 ± 0.09 d	10.90 ± 1.24 c	8.2
LoUS.Psy_25.75	29.98 ± 0.82 d	−0.06 ± 0.23 d	14.10 ± 0.46 b,c	5.5
LoUS.Psy_50.50	26.96 ± 1.17 d,e	−0.40 ± 0.09 d	12.38 ± 1.15 c	6.9
LoUS.Psy_75.25	25.87 ± 0.55 e	−0.64 ± 0.06 d	11.60 ± 1.14 c	7.7

Psy (control), *Psyllium* gels prepared in water; PsyL, *Psyllium* gels prepared in autohydrolysis liquor; *Laminaria-Psyllium* gels: Lo.Psy_25.75, Lo.Psy_50.50, and LoPsy_75.25; *Laminaria* with ultrasonic treatment-*Psyllium* gels: LoUS.Psy_25.75, LoUS.Psy_50.50, and LoUSPsy_75.25. Data are presented as the mean ± sd. Different letters in the same column correspond to significant differences ($p < 0.05$).

As expected, the lightness (L*) of the gel samples decreased with the incorporation of *Laminaria*, although significant differences were only registered between the *Laminaria-Psyllium* 25.75 and *Laminaria-Psyllium* 75.25 samples. It is noteworthy that the color parameters of the gel samples with the highest *Laminaria* content were like those obtained in the purée-like mixtures [14]. Moreover, the color parameters L* (Lo.Psy: r = −0.928, $p = 1.3 \times 10^{-9}$; LoUS.Psy: r = −0.920, $p = 9.3 \times 10^{-9}$) and a* (Lo.Psy: r = −0.849; $p = 1.2 \times 10^{-6}$; LoUS.Psy: r = −0.913, $p = 2.0 \times 10^{-8}$) were strongly negatively correlated with *Laminaria* content in the gel formulations.

The ΔE* between the *Laminaria-Psyllium* gels and the control ranged from 4.5 to 8.2, increasing with the increase in the *Laminaria* content in the system. These values indicate that the consumer would distinguish between the two samples compared [25].

It should be pointed out that the gels proposed here are not finished products. In this sense, the color of the final food gel product could be optimized by considering these findings and according to the desired final color.

3.3. Effect of Laminaria-Psyllium Ratio on the Dynamic Viscoelasticity

Figure 2 shows the elastic behavior of *Laminaria-Psyllium* gel systems. As the alga fraction increased, a structuring effect was observed in the gels. This behavior was markedly noticed in the highest alga concentration (Lo.Psy_75.25 and LoUS.Psy_75.25). This interaction is probably due to physical entanglements between the polymers present in both the alga and *Psyllium*, causing the reinforcement of the gel. However, in the absence of the *Psyllium* gel, the alga purées showed a huge decrease in G′ to values similar to the control (Psy), but more frequency dependent (b′ = 0.209) [17].

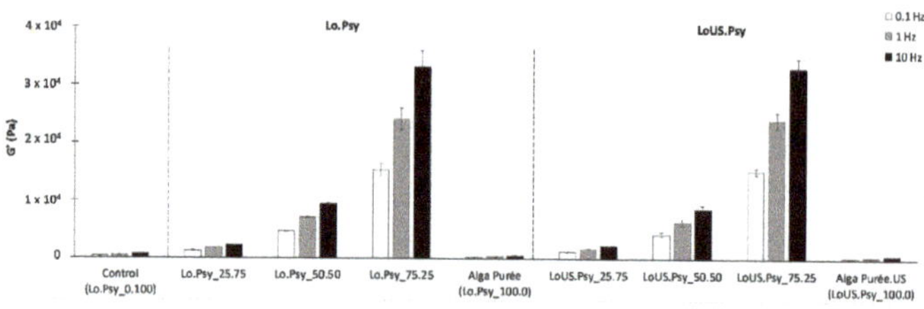

Figure 2. Elastic modulus at 0.1 Hz, 1 Hz, and 10 Hz of *Laminaria-Psyllium* gels with (LoUS.Psy) and without ultrasonic treatment (Lo.Psy), control, and *Laminaria* purées (Alga Purée; Alga Purée.US) [17].

This synergistic interaction was also found in other polymeric systems, namely between the locust bean gum (LBG) and xanthan gum, where the latter did not form gel, rather a shear-thinning solution, but combined with LBG to form a strong gel structure [26].

The mechanical spectra of the developed gels are depicted in Figure 3.

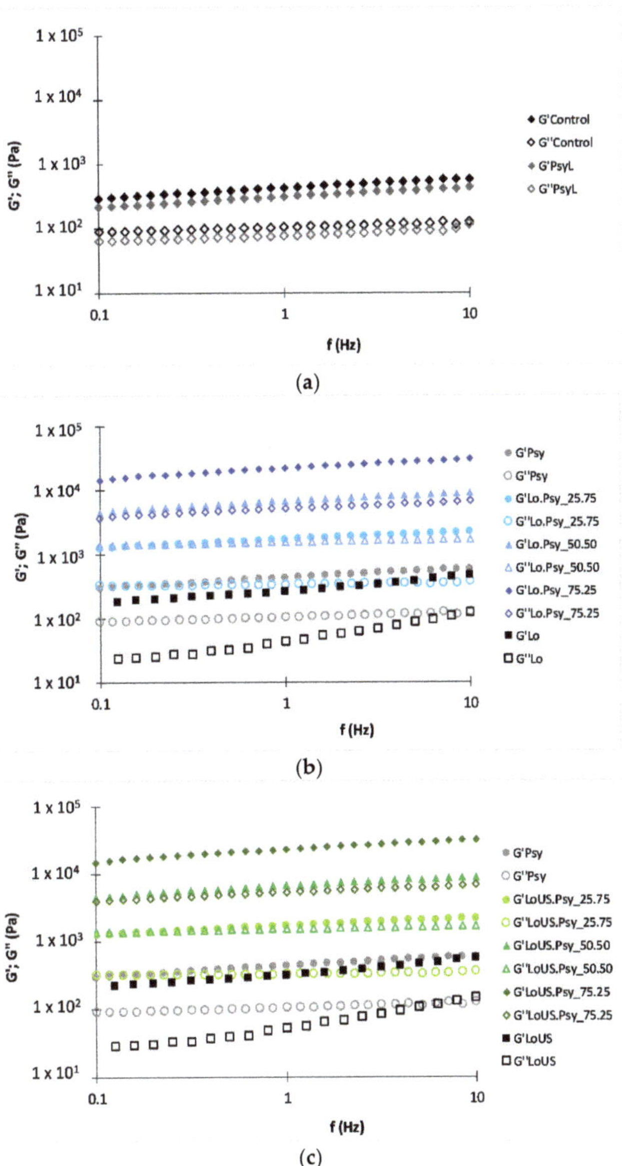

Figure 3. Mechanical spectra of the *Psyllium* gels prepared in water (control) and autohydrolysis liquor (PsyL). (**a**) *Laminaria-Psyllium* gels (Lo.Psy_25.75, Lo.Psy_50.50, LoPsy_75.25); (**b**) *Laminaria* with ultrasonic treatment-*Psyllium* gels (LoUS.Psy_25.75, LoUS.Psy_50.50, LoUSPsy_75.25; (**c**) G′, closed symbol; G″, open symbol.

To quantify the impact of the different combinations of *Laminaria* and *Psyllium* on the viscoelastic moduli, the variation of G′ and G″ with gel composition was obtained from the respective mechanical spectra (Table 3).

Table 3. Power law parameters (α', α'', b′, and b″) of the gel samples with *Laminaria ochroleuca* and *Psyllium* husk, control, and PsyL.

Samples	G′		G″	
	α'	b′	α''	b″
Psy (Control)	436.8 ± 15.4	0.153 ± 0.002	104.8 ± 7.0	0.071 ± 0.011
PsyL	323.0 ± 18.2	0.149 ± 0.005	79.6 ± 2.4	0.090 ± 0.005
Lo.Psy_25.75	1689.3 ± 20.8	0.124 ± 0.000	344.2 ± 2.7	0.022 ± 0.000
Lo.Psy_50.50	6819.9 ± 103.8	0.148 ± 0.002	1584.5 ± 28.3	0.049 ± 0.002
Lo.Psy_75.25	0.5 ± 1816.6	0.158 ± 0.002	5485.0 ± 434.5	0.119 ± 0.001
LoUS.Psy_25.75	1681.6 ± 3.5	0.121 ± 0.000	331.9 ± 2.8	0.019 ± 0.000
LoUS.Psy_50.50	6239.3 ± 543.3	0.146 ± 0.002	1442.5 ± 122.0	0.049 ± 0.001
LoUS.Psy_75.25	23,533.5 ± 1267.8	0.159 ± 0.001	5530.1 ± 331.8	0.116 ± 0.002

The goodness of fitting (R^2) ranged from 0.995–0.997 for G′ and from 0.936–0.997 for G″.

The mechanical spectra Psy (control) and PsyL exhibited similar viscoelastic performance typical of well-structured weak gels, with slight frequency dependence (Figure 3a). This result is consistent with the rheological study by Haque et al. [27], in which it is also reported that *Psyllium* husk forms gel even at low temperature, this being an important feature to consider in food design. Moreover, the developed gels were more stable at higher frequencies than the ones produced with 10–15% chia flour at 90 °C (Ramos et al., 2016), reinforcing *Psyllium*'s potential as a valuable and sustainable biopolymer.

As mentioned earlier, PsyL appeared to be more fluid than the control (Psy), which was confirmed by the mechanical spectra, with both viscoelastic moduli of the control being higher than those of PsyL, and by the decrease of α' (Table 3). Autohydrolysis promotes the solubilization of minerals [17] and the depolymerization of polysaccharides, namely alginate, fucoidan, and laminarin present in brown algae [28], rendering a liquid extract with an acidic pH ($\approx$5, [19]). Since *Laminaria* liquid extract is a multicomponent matrix, its effect on the *Psyllium* gel properties are more complex, depending not only on the solution pH, but also on the type of ions in the solution, and even on the presence of peptides and other molecules resulting from the depolymerization of the polysaccharides. These polymer fractions can interfere with the gel matrix and exert an antagonism, leading to the reduction of gel links and de-structuring the material.

The ultrasonic pre-treatment applied to *L. ochroleuca* did not affect the rheological behavior of the *Laminaria-Psyllium* gels. The weak-gel like behavior was maintained in all *Laminaria-Psyllium* gel samples, and was more noticeable as the alga fraction increased, as can be observed by the increase in the b′ parameter of the resulting power law (Table 3). As the incorporation of the alga solid fraction increased, the dependence of the material on frequency increased; however the value of the viscoelastic moduli increased probably due to the reduction of the number of links, but also stronger ones, which may be due to interactions between the alginate from the alga and fiber from the *Psyllium* husk reinforced by calcium ions [29].

3.4. Texture Properties of the Laminaria-Psyllium Gel Systems

Texture is a major quality parameter, and is crucial for consumer acceptance. Figure 4 presents the texture profiles of the samples and the commercial references.

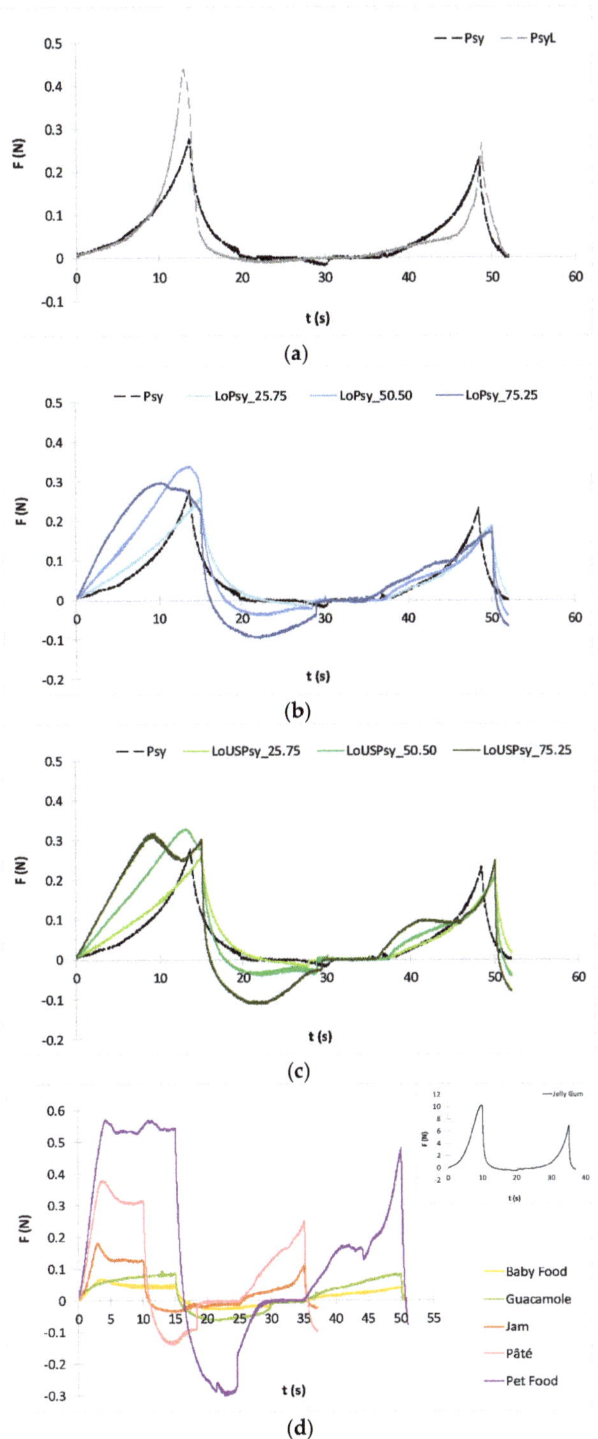

Figure 4. Texture profiles of Psy (control) and PsyL (**a**), *Laminaria-Psyllium* gels (**b**), *Laminaria-Psyllium* gels subjected to ultrasonic treatment (**c**), and commercial products (**d**). The differences in the texturograms reflect the different sizes and shapes of the commercial products.

Each food system (e.g., jelly, mayonnaise, and baby food) presents texture properties that are specific, easily recognizable, and desired by the consumers. Despite the information given by the texture profile, the use of this methodology is very limited in the scientific literature. To our knowledge, no published studies refer to the texture features of commercial gelled products, therefore we show here some texture profiles of different commercial food products to illustrate the data for these systems to be used as a target for further product development. Based on these texture profiles (Figure 4d), three groups of samples can be discriminated: one composed of *Pâté*, *Pet Food* and *Jam*; another with *Baby Food* and *Guacamole*; and a third with *Jelly Gum*, whose graph is shown separately. In the first group of samples, the rupture point occurred at a low break distance, and the force reached a plateau until the probe retracted from the sample. This mechanical behavior was similar to what we found in LoUS.Psy_75.25 (Figure 4c) and has also been reported by Genovese and co-workers [30] for pectin gels. The control and PsyL texture profiles (Figure 4a) were similar to the one obtained by Figueroa et al. [21] for fruit jellies with *Psyllium*, although with much lower magnitude.

From the texture profiles depicted, the firmness, adhesiveness, and cohesiveness values were calculated (Table 4).

Table 4. Texture parameters (firmness, adhesiveness and cohesiveness) of the gels developed and commercial references.

	Samples	Firmness (N)	Adhesiveness (N.s)	Cohesiveness
	Psy (control)	0.296 ± 0.036 c,d,e	−0.034 ± 0.006 a	0.581 ± 0.051 b
	PsyL	0.419 ± 0.031 b,c	−0.032 ± 0.003 a	0.453 ± 0.024 c,d,e
Developed gel samples	**Lo.Psy_25.75**	0.259 ± 0.028 c,d,e	−0.040 ± 0.012 a	0.517 ± 0.037 b,c
	Lo.Psy_50.50	0.316 ± 0.020 c,d,e	−0.274 ± 0.038 a	0.362 ± 0.044 e,f,g
	Lo.Psy_75.25	0.279 ± 0.039 c,d,e	−0.977 ± 0.075 c	0.403 ± 0.026 d,e,f
	LoUS.Psy_25.75	0.261 ± 0.007 c,d,e	−0.069 ± 0.017 a	0.538 ± 0.009 b,c
	LoUS.Psy_50.50	0.333 ± 0.026 b,c,d	−0.268 ± 0.042 a	0.354 ± 0.028 f,g
	LoUS.Psy_75.25	0.319 ± 0.024 c,d,e	−0.978 ± 0.047 c	0.391 ± 0.030 d,e,f,g
	Baby Food	0.065 ± 0.008 d,e	−0.325 ± 0.034 a,b	0.475 ± 0.023 c,d
	Jelly Gum	10.314 ± 0.386 a	−3.596 ± 0.562 e	0.330 ± 0.028 f,g
Commercial products	**Guacamole**	0.091 ± 0.011 e	−0.638 ± 0.106 b,c	0.748 ± 0.075 a
	Jam	0.182 ± 0.003 c,d,e	−0.298 ± 0.015 a	0.310 ± 0.015 g
	Pâté	0.364 ± 0.032 b,c	−0.854 ± 0.081 c	0.464 ± 0.038 c,d
	Pet Food	0.587 ± 0.066 b	−2.173 ± 0.148 d	0.371 ± 0.024 e,f,g

Psy (control), *Psyllium* gels prepared in water; PsyL, *Psyllium* gels prepared in autohydrolysis liquor; *Laminaria-Psyllium* gels: Lo.Psy_25.75, Lo.Psy_50.50, and LoPsy_75.25; *Laminaria* with ultrasonic treatment-*Psyllium* gels: LoUS.Psy_25.75, LoUS.Psy_50.50, and LoUSPsy_75.25. Data are presented as the mean ± sd. Different letters in the same column correspond to significant differences ($p < 0.05$).

In general, it can be said that the firmness of the *Laminaria-Psyllium* system is independent of the level of *Laminaria* incorporation and the ultrasonic (US) pre-treatment. On the other hand, there are strong negative correlations between the adhesiveness (r = −0.888, $p < 0.05$) and cohesiveness (r = −0.844, $p < 0.05$) parameters and *Laminaria* concentration, and again, the US pre-treatment did not make any change. The fact that adhesiveness can be adjusted by keeping the firmness values constant could be useful in product development.

In a previous study by the authors [17], it was found that the texture parameters, especially adhesiveness, of the alga purée-like mixtures varied according to the ultrasonic treatment. However, that difference was not maintained after the addition of *Psyllium*. Since PsyL did not show adhesiveness values that could explain this result, one can conclude that there is a synergistic effect of *Laminaria* and *Psyllium* in this texture parameter, stronger than mere addition. Based on these results and looking at the texture profiles of the samples (Figure 4a–c), one can conclude that the texture properties of *Laminaria-Psyllium* gels are governed by the alga, what is in agreement with the dynamic rheology results. Texture profiles (Figure 4) confirm the relevance of alga in the system.

The texture parameters of the commercial products varied greatly. Comparing the texture parameters of the gelled systems developed with those of the references (Table 4), the developed *Laminaria-Psyllium* gels presented similar values to the commercial products. Although the systems developed are not finished products, but binary systems consisting of alga purée and *Psyllium* gel, this is a good starting point for the development of enriched gelled food products. These *Laminaria* and *Psyllium* whole materials can be used as alternatives to the biopolymers that are usually added to build up structure in foods. This is in line with the food trends regarding the use of natural products over processed ones.

It should be pointed out that in this approach, we developed the *Laminaria-Psyllium* gels using water, but the gels could also be produced in the same way using the AH liquor extract, thus taking advantage of the soluble compounds present there, namely phenolic compounds with antioxidant capacity [17]. Due to the differences in texture and rheological features between the control and PsyL, it is plausible to assume that *Laminaria-Psyllium* gels prepared in AH liquor would present distinct mechanical properties.

4. Conclusions

Food gels were developed with *Laminaria* and *Psyllium* husk, adding value to both these natural resources. *Psyllium* husk interacted positively with this alga, reinforcing the viscoelastic behavior of the obtained gels. These novel functional gels with minimal processing can be a starting point to the development of several food and feed applications.

Food application studies are in progress to evaluate the demonstrated potential of this *Laminaria-Psyllium* partnership.

Author Contributions: P.F., M.D.T., and A.R. conceived and planned the experiments. P.F. and M.D.T. participated in the sample preparation and analysis, data analysis and interpretation of the results, and writing of the manuscript with input from all authors. I.S., A.R., and H.D. supervised the research work, contributed to the discussion of the data, and revised the manuscript. All authors contributed with suggestions and comments for the final version of the manuscript.

Funding: The authors are grateful to the Spanish Ministry of Economy and Competitiveness CTM2015-68503-R). M.D.T. thanks the Spanish Ministry of Economy and Competitiveness for her postdoctoral grant (IJCI-2016-27535). P.F. acknowledges her PhD grant from the Universidade de Lisboa (C0144M) and ERASMUS+ grant. The authors are also grateful to Fundação para a Ciência e a Tecnologia (Portugal) through the research unit UID/AGR/04129/2013 (LEAF) financial support.

Conflicts of Interest: The authors declare no conflict of interest.

References

1. Kim, S.-K.; Bhatnagar, I. Physical, Chemical, and Biological Properties of Wonder Kelp—*Laminaria*. In *Advances in Food and Nutrition Research*; Kim, S.-K., Ed.; Elsevier Inc.: Amsterdam, The Netherlands, 2011; pp. 85–96. [CrossRef]
2. Adebisi, A.O.; Laity, P.R.; Conway, B.R. Formulation and evaluation of floating mucoadhesive alginate beads for targeting *Helicobacter pylori*. *J. Pharm. Pharmacol.* **2015**, *67*, 511–524. [CrossRef] [PubMed]
3. Oliveira, S.; Fradinho, P.; Mata, P.; Moreira-Leite, B.; Raymundo, A. Exploring innovation in a traditional sweet pastry: Pastel de Nata. *Int. J. Gastron. Food Sci.* **2019**, *17*, 100160. [CrossRef]
4. Sánchez-Machado, D.I.; López-Cervantes, J.; López-Hernández, J.; Paseiro-Losada, P. Fatty acids, total lipid, protein and ash contents of processed edible seaweeds. *Food Chem.* **2004**, *85*, 439–444. [CrossRef]
5. Fernandes, F.; Barbosa, M.; Oliveira, A.P.; Azevedo, I.C.; Sousa-Pinto, I.; Valentão, P.; Andrade, P.B. The pigments of kelps (*Ochrophyta*) as part of the flexible response to highly variable marine environments. *J. Appl. Phycol.* **2016**, *28*, 3689–3696. [CrossRef]
6. Parada, J.; Pérez-Correa, J.R.; Pérez-Jiménez, J. Design of low glycemic response foods using polyphenols from seaweed. *J. Funct. Foods.* **2019**, *56*, 33–39. [CrossRef]
7. Andrade, P.B.; Barbosa, M.; Matos, R.P.; Lopes, G.; Vinholes, J.; Mouga, T.; Valentão, P. Valuable compounds in macroalgae extracts. *Food Chem.* **2013**, *138*, 1819–1828. [CrossRef]

8. Roohinejad, S.; Koubaa, M.; Barba, F.J.; Saljoughian, S.; Amid, M.; Greiner, M. Application of seaweeds to develop new food products with enhanced shelf-life, quality and health-related beneficial properties. *Food Res. Int.* **2017**, *99*, 1066–1083. [CrossRef]

9. FDA. CFR–code of federal regulations title 21. U.S. Food and Drug Administration. 2012. Available online: http://www.accessdata.fda.gov/scripts/cdrh/cfdocs/cfcfr/CFRSearch.cfm?fr=101.81 (accessed on 26 July 2019).

10. Raymundo, A.; Fradinho, P.; Nunes, M.C. Effect of *Psyllium* fibre content on the textural and rheological characteristics of biscuit and biscuit dough. *Bioact. Carbohydr. Dietary Fibre.* **2014**, *3*, 96–105. [CrossRef]

11. Fischer, M.H.; Yu, N.; Gray, G.R.; Ralph, J.; Anderson, L.; Marlett, J.A. The gel-forming polysaccharide of psyllium husk (*Plantago ovata* Forsk). *Carbohyd. Res.* **2004**, *339*, 2009–2017. [CrossRef]

12. Askari, F.; Sadeghi, E.; Mohammadi, R.; Rouhi, M.; Taghizadeh, M.; Shirgardoun, M.H.; Kariminejad, M. The physicochemical and structural properties of psyllium gum/modified starch composite edible film. *J. Food Process Preserv.* **2018**, *42*, 13715. [CrossRef]

13. Singh, B. Psyllium as therapeutic and drug delivery agent. *Int. J. Pharm.* **2007**, *334*, 1–14. [CrossRef] [PubMed]

14. Ramos, S.; Fradinho, P.; Mata, P.; Raymundo, A. Assessing gelling properties of chia (*Salvia hispanica L.*) flour through rheological characterization. *J. Sci. Food Agric.* **2016**, *97*, 1753–1760. [CrossRef] [PubMed]

15. Soukoulis, C.; Cambier, S.; Serchi, T.; Tsevdou, M.; Gaiani, C.; Ferrer, P.; Taoukis, P.S.; Hoffmann, L. Rheological and structural characterisation of whey protein acid gels co-structured with chia (*Salvia hispanica L.*) or flax seed (*Linum usitatissimum L.*) mucilage. *Food Hydrocolloid.* **2019**, *89*, 542–553. [CrossRef]

16. Soukoulis, C.; Gaiani, C.; Hoffmann, L. Plant seed mucilage as emerging biopolymer in food industry applications. *Curr. Opin. Food Sci.* **2018**, *22*, 28–42. [CrossRef]

17. Fradinho, P.; Flórez-Fernández, N.; Sousa, I.; Raymundo, A.; Domínguez, H.; Torres Pérez, M.D. Environmentally friendly processing of *Laminaria ochroleuca* for soft food applications with bioactive properties. *J. Appl. Phycol.* **2019**. accepted for publication.

18. Fradinho, P.; Sousa, I.; Raymundo, A. The role of *Psyllium* gels on the structuring of gluten-free fresh pasta –a rheological approach. In *Book of Abstracts of AERC 2019. Proceedings of the Oral Communication FP11, Portoröz, Slovenia, 8–11 April 2019*; The European Society of Rheology: Zurich, Switzerland, 2019; p. 29.

19. Gómez-Ordóñez, E.; Jiménez-Escrig, A.; Rupérez, P. Dietary fibre and physicochemical properties of several edible seaweeds from the northwestern Spanish coast. *Food Res. Int.* **2010**, *43*, 2289–2294. [CrossRef]

20. Paraskevopoulou, A.; Kiosseoglou, V.; Alevisopoulos, S.; Kasapis, S. Small deformation properties of model salad dressings prepared with reduced cholesterol egg yolk. *J. Texture Stud.* **1997**, *28*, 221–237. [CrossRef]

21. Figueroa, L.E.; Genovese, D.B. Fruit jellies enriched with dietary fibre: Development and characterization of a novel functional food product. *LWT Food Sci. Technol.* **2019**, *111*, 423–428. [CrossRef]

22. Flórez-Fernández, N.; Torres, M.D.; González-Muñoz, M.J.; Domínguez, H. Recovery of bioactive and gelling extracts from edible brown seaweed *Laminaria ochroleuca* by non-isothermal autohydrolysis. *Food Chem.* **2019**, *277*, 353–361. [CrossRef]

23. Margier, M.; Georgé, S.; Hafnaoui, N.; Remond, D.; Nowicki, M.; Du Chaffaut, L.; Amiot, M.-J.; Reboul, E. Nutritional Composition and Bioactive Content of Legumes: Characterization of Pulses Frequently Consumed in France and Effect of the Cooking Method. *Nutrients* **2018**, *10*, 1668. [CrossRef]

24. Jimoh, M.A.; MacNaughtan, W.; Williams, H.E.L.; Greetham, D.; Linforth, R.L.; Fisk, I.D. Sodium ion interaction with psyllium husk (*Plantago sp.*). *Food Funct.* **2016**, *7*, 4041. [CrossRef] [PubMed]

25. Castellar, M.R.; Obón, J.M.; Fernández-López, J.A. The isolation and properties of a concentrated red-purple betacyanin food colorant from *Opuntia stricta* fruits. *J. Sci. Food Agric.* **2006**, *86*, 122–128. [CrossRef]

26. Copetti, G.; Grassi, M.; Lapasin, R.; Pricl, S. Synergistic gelation of xanthan gum with locust bean gum: A rheological investigation. *Glycoconj. J.* **1997**, *14*, 951–961. [CrossRef] [PubMed]

27. Haque, A.; Richardson, R.K.; Morris, E.R.; Dea, I.C.M. Xanthan-like 'weak gel' rheology from dispersions of ispaghula seed husk. *Carbohyd. Polym.* **1993**, *22*, 223–232. [CrossRef]

28. Rodríguez-Jasso, R.M.; Mussatto, S.I.; Pastrana, L.; Aguilar, C.N.; Teixeira, J.A. Extraction of sulfated polysaccharides by autohydrolysis of brown seaweed *Fucus vesiculosus*. *J. Appl. Phycol.* **2013**, *25*, 31–39. [CrossRef]

29. Guo, Q.; Cui, S.W.; Wang, Q.; Goff, H.D.; Smith, A. Microstructure and rheological properties of psyllium polysaccharide gel. *Food Hydrocolloid.* **2009**, *23*, 1542–1547. [CrossRef]
30. Genovese, D.B.; Ye, A.; Singh, H. High Methoxyl Pectin/Apple Particles Composite Gels: Effect of Particle Size and Particle concentration on Mechanical Properties and Gel Structure. *J. Text Stud.* **2010**, *41*, 171–189. [CrossRef]

MDPI

St. Alban-Anlage 66

4052 Basel

Switzerland

Tel. +41 61 683 77 34

Fax +41 61 302 89 18

www.mdpi.com

Applied Sciences Editorial Office

E-mail: applsci@mdpi.com

www.mdpi.com/journal/applsci

www.ingramcontent.com/pod-product-compliance
Lightning Source LLC
LaVergne TN
LVHW070620170726
843515LV00004B/138